T0351315

Structural and Chemical Characterization of Metals, Alloys, and Compounds—2014

MATERIALS RESEARCH SOCIETY
SYMPOSIUM PROCEEDINGS VOLUME 1766

Structural and Chemical Characterization of Metals, Alloys, and Compounds—2014

Symposium held August 17-21, 2014, Cancún, México

EDITORS

Ramiro Pérez Campos

Centro de Física Aplicada y Tecnología Avanzada, UNAM
Querétaro, México

Antonio Contreras Cuevas

Instituto Mexicano del Petróleo
San Bartolo Atepehuacan, México

Rodrigo A. Esparza Muñoz

Centro de Física Aplicada y Tecnología Avanzada, UNAM
Querétaro, México

Materials Research Society
Warrendale, Pennsylvania

CAMBRIDGE
UNIVERSITY PRESS

Shaftesbury Road, Cambridge CB2 8EA, United Kingdom

One Liberty Plaza, 20th Floor, New York, NY 10006, USA

477 Williamstown Road, Port Melbourne, VIC 3207, Australia

314–321, 3rd Floor, Plot 3, Splendor Forum, Jasola District Centre, New Delhi – 110025, India

103 Penang Road, #05–06/07, Visioncrest Commercial, Singapore 238467

Cambridge University Press is part of Cambridge University Press & Assessment, a department of the University of Cambridge.

We share the University's mission to contribute to society through the pursuit of education, learning and research at the highest international levels of excellence.

www.cambridge.org
Information on this title: www.cambridge.org/9781605117430

Materials Research Society
506 Keystone Drive, Warrendale, PA 15086
http://www.mrs.org

First published 2015

CODEN: MRSPDH

A catalogue record for this publication is available from the British Library

ISBN   978-1-605-11743-0   Hardback

CONTENTS

v

CHARACTERIZATION OF STEELS USED IN THE OIL INDUSTRY

CHARACTERIZATION OF MATERIALS FOR INDUSTRIAL APPLICATIONS

CHARACTERIZATION OF MATERIALS USED IN COATINGS AND THIN FILMS

CHARACTERIZATION OF NANOSTRUCTURED MATERIALS

PREFACE

The XXIII International Materials Research Congress was held on August 17–21, 2014, in Cancún, Mexico. It was organized by the Sociedad Mexicana de Materiales (SMM). About 1,500 specialized scientists from more than 40 countries participated in the 30 different symposia, workshops, plenary lectures and tutorial courses. The 30 symposia that comprise the technical program of IMRC 2014 are grouped in several clusters, namely: Nanoscience and Nanotechnology, Biomaterials, Materials for Energy, Fundamental Materials Science, Materials Characterization, Materials for Specific Applications, Magnetic and Electronic Materials and General.

This Materials Research Society Proceedings volume contains papers presented at the Symposium 5B "Structural and Chemical Characterization of Metals, Alloys and Compounds" of the XXIII International Materials Research Congress. This event is intended to be a forum for the dissemination of research results on materials research. The participants and the organizers have found this event to be very successful due to the high quality and novelty of the scientific results presented. Among the important achievements of the symposium are the new personal contacts among the scientists for the creation of multinational thematic and research networks, as well as promoting contacts for future collaboration.

This special issue covers several aspects of the structural and chemical characterization of materials in the following areas: metals, alloys, steels, composites, polymeric compounds, welding, nanomaterials, and surface coatings, among others. They are amorphous, crystalline, powders, coatings, fibers, thin films, and so forth, which were prepared with different techniques. The structural characterization techniques include: scanning electron microscopy (SEM), X-ray diffraction (XRD), transmission electron microscopy (TEM), Raman spectroscopy, optical microscopy (OM), Fourier transform infrared spectroscopy (FTIR), differential thermal analysis (DTA), differential scanning calorimetry (DSC), thermogravimetry analysis (TGA), thermo luminescence (TL), laser emission, and so forth. Theoretical models from these properties are included too.

The scientific program of Symposium 5B includes 69 oral and 144 poster presentations. In addition, this year the invited talks were focused on X-ray diffraction technique applied to the characterization of materials. This special issue contains 20 papers based on contributions presented during the symposium. All manuscripts included in this special issue have been accepted after peer review.

Dr. Ramiro Pérez Campos
Dr. Antonio Contreras Cuevas
Dr. Rodrigo A. Esparza Muñoz

December, 2014

Acknowledgments

We would like to thank the members of the MRS-Mexico Advisory Committee and the reviewers for their valuable comments, which have certainly helped to improve the quality of the manuscripts. We also wish to thank the Sociedad Mexicana de Materiales (SMM), Universidad Nacional Autonoma de México (UNAM) and Instituto Mexicano del Petróleo (IMP) for their support in organizing Symposium 5B "Structural and chemical characterization of metals, alloys and compounds."

Additionally, we would like to thank those who have worked to make this congress an exciting and fruitful meeting: meeting chairs, symposia organizers, IMRC staff, MRS staff, editors, management committee, advisory committee, and Sociedad Mexicana de Materiales (SMM).

MATERIALS RESEARCH SOCIETY SYMPOSIUM PROCEEDINGS

Volume 1717E–   Organic Bioelectronics, 2015, M.R. Abidian, C. Bettinger, R. Owens, D.T. Simon, ISBN 978-1-60511-694-5

Volume 1718–   Multifunctional Polymeric and Hybrid Materials, 2015, A. Lendlein, N. Tirelli, R.A. Weiss, T. Xie, ISBN 978-1-60511-695-2

Volume 1719E–   Medical Applications of Noble Metal Nanoparticles (NMNPs), 2015, X. Chen, H. Duan, Z. Nie, H-R. Tseng, ISBN 978-1-60511-696-9

Volume 1720E–   Materials and Concepts for Biomedical Sensing, 2015, X. Fan, L. Liu, E. Park, H. Schmidt, ISBN 978-1-60511-697-6

Volume 1721E – Hard-Soft Interfaces in Biological and Bioinspired Materials—Bridging the Gap between Theory and Experiment, 2015, J. Harding, D. Joester, R. Kröger, P. Raiteri, ISBN 978-1-60511-698-3

Volume 1722E–   Reverse Engineering of Bioinspired Nanomaterials, 2015, L. Estroff, S-W. Lee, J-M. Nam, E. Perkins, ISBN 978-1-60511-699-0

Volume 1723E–   Plasma Processing and Diagnostics for Life Sciences, 2015, E.R. Fisher, M. Kong, M. Shiratani, K.D. Weltmann, ISBN 978-1-60511-700-3

Volume 1724E–   Micro/Nano Engineering and Devices for Molecular and Cellular Manipulation, Simulation and Analysis, 2015, D.L. Fan, J. Fu, X. Jiang, M. Lutolf, ISBN 978-1-60511-701-0

Volume 1725E–   Emerging 1D and 2D Nanomaterials in Health Care, 2015, P.M. Ajayan, S.J. Koester, M.R. McDevitt, V. Renugopalakrishnan, ISBN 978-1-60511-702-7

Volume 1726E–   Emerging Non-Graphene 2D Atomic Layers and van der Waals Solids, 2015, M. Bar-Sadan, J. Cheon, S. Kar, M. Terrones, ISBN 978-1-60511-703-4

Volume 1727E–   Graphene and Graphene Nanocomposites, 2015, J. Jasinski, H. Ji, Y. Zhu, V. Nicolosi, ISBN 978-1-60511-704-1

Volume 1728E–   Optical Metamaterials and Novel Optical Phenomena Based on Nanofabricated Structures, 2015, Y. Liu, F. Capasso, A. Alú, M. Agio, ISBN 978-1-60511-705-8

Volume 1729–   Materials and Technology for Nonvolatile Memories, 2015, P. Dimitrakis, Y. Fujisaki, G. Hu, E. Tokumitsu, ISBN 978-1-60511-706-5

Volume 1730E–   Frontiers in Complex Oxides, 2015, J.D. Baniecki, N.A. Benedek, G. Catalan, J.E. Spanier, ISBN 978-1-60511-707-2

Volume 1731E–   Oxide semiconductors, 2015, T.D. Veal, O. Bierwagen, M. Higashiwaki, A. Janotti, ISBN 978-1-60511-708-9

Volume 1732E–   Hybrid Oxide/Organic Interfaces in Organic Electronics, 2015, A. Amassian, J.J. Berry, M.A. McLachlan, E.L. Ratcliff, ISBN 978-1-60511-709-6

Volume 1733E–   Fundamentals of Organic Semiconductors—Synthesis, Morphology, Devices and Theory, 2015, D. Seferos, L. Kozycz, ISBN 978-1-60511-710-2

Volume 1734E–   Diamond Electronics and Biotechnology—Fundamentals to Applications, 2015, C-L. Cheng, D.A.J. Moran, R.J. Nemanich, G.M. Swain, ISBN 978-1-60511-711-9

Volume 1735–   Advanced Materials for Photovoltaic, Fuel Cell and Electrolyzer, and Thermoelectric Energy Conversion, 2015, S.R. Bishop, D. Cahen, R. Chen, E. Fabbri, F.C. Fonseca, D. Ginley, A. Hagfeldt, S. Lee, J. Liu, D. Mitzi, T. Mori, K. Nielsch, Z. Ren, P. Rodriguez, ISBN 978-1-60511-712-6

Volume 1736E–   Wide-Bandgap Materials for Solid-State Lighting and Power Electronics, 2015, R. Kaplar, G. Meneghesso, B. Ozpineci, T. Takeuchi, ISBN 978-1-60511-713-3

Volume 1737E–   Organic Photovoltaics—Fundamentals, Materials and Devices, 2015, A. Facchetti, ISBN 978-1-60511-714-0

Volume 1738E–   Sustainable Solar-Energy Conversion Using Earth-Abundant Materials, 2015, Y. Li, S. Mathur, G. Zheng, ISBN 978-1-60511-715-7

Volume 1739E–   Technologies for Grid-Scale Energy Storage, 2015, B. Chalamala, J. Lemmon, V. Subramanian, Z. Wen, ISBN 978-1-60511-716-4

Volume 1740E–   Materials Challenges for Energy Storage across Multiple Scales, 2015, A. Cresce, ISBN 978-1-60511-717-1

Volume 1741E   Synthesis, Processing and Mechanical Properties of Functional Hexagonal Materials, 2015, M. Albrecht, S. Aubry, R. Collazo, R.K. Mishra, C-C. Wu, ISBN 978-1-60511-718-8

Volume 1742E–   Molecular, Polymer and Hybrid Materials for Thermoelectrics, 2015, A. Carella, M. Chabinyc, M. Kovalenko, J. Malen, R. Segalman, ISBN 978-1-60511-719-5

Volume 1743E–   Materials and Radiation Effects for Advanced Nuclear Technologies, 2015, G. Baldinozzi, C. Deo, K. Arakawa, F. Djurabekova, S.K. Gill, E. Marquis, F. Soisson, K. Yasuda, Y. Zhang, ISBN 978-1-60511-720-1

MATERIALS RESEARCH SOCIETY SYMPOSIUM PROCEEDINGS

Prior Materials Research Symposium Proceedings available by contacting Materials Research Society

Characterization of Materials for Medical Applications

Mater. Res. Soc. Symp. Proc. Vol. 1766 © 2015 Materials Research Society
DOI: 10.1557/opl.2015.406

Study Microstructure and Mechanical Properties of Prosthesis of Forging

D. C. Rojas-Olmos[1], N. López-Perrusquia[1], M. A. Doñu-Ruiz[1], J.A Juanico Loran[1], C. R. Torres San Miguel [2]

[1]Universidad Politécnica Valle de México; *Grupo Ciencia e Ingeniería de Materiales*, UPVM, Tultitlán. Edo de México.
E-mail: nocperrusquia@hotmail.com
[2]Instituto Politécnico Nacional, SEPI-Esime, Adolfo López Mateos, Zacatenco, México D.F, C. P. 07738, México.

ABSTRACT

This work studies the change microstructural and mechanical properties of biomedical component hot forging of titanium; was assessed quantitatively and qualitatively the microstructural features obtained in this titanium biocompatible Ti6Al4V. The forging process was obtained at temperature of 950 °C, after by technical optical microscopy are obtained the microstructural characterization showing the phases present after forging. Likewise, the technical X-ray diffraction (XRD) shows the presence of the phases. Also is evaluated the hardness and modulus of elasticity by technical nanoindentation. The characterization of this material has the objective to show that the results obtained with temperature study of 950 °C . Likewise by the forging process obtained a type phases and optimal properties required for these biomedical materials.

INTRODUCTION

The industry of biomedical implants internal and external has become the last few decades in one of the most dynamic and growing annual; this trend should continue to respond to needs of the global society; this with the aim of developing implants that give a best possible quality of life [1]. The titanium alloys Ti–6Al–4V are metallic materials for biomedical applications [2 -3].

The evolution of this manufacturing industry of biomedical prosthesis has been greatly enhanced by the studies of the materials used in this area of health [4]. The characterization of deformation behavior is thus essential for the optimization of hot forging processes of titanium alloys [5-6]. The manufacturing processes are an important part in the production of biomedical implants [7]. In the forging process can obtain phase α and β [8]; this have greater biocompatibility phase β types [9]. The mechanical properties of the biomedical implants, after processes of manufactures are important by changes microstructure and hot workability of the forging [10]. The decisive innovations in manufacturing of these biomedical components, by other manufacturing techniques are productive and more efficient [11]. In addition the training of biomedical implants by the manufacture of wrought iron, is a new technology.

The forging of titanium alloy blades is a difficult operation to describe quantitatively due to the changing properties of the workpiece material during forging process [12-13].

EXPERIMENTAL

Used a lubricant for hot forging test, based glass, the press speed was 16 mm s−1.The piece of forged alloy Ti6Al4V is shown in Figure 1. The chemical composition is shown in Table I. Microstructural characterization was determined by means of a metallographic microscope Olympus GX 51 along the ankle prosthesis, also by scanning electron microscopy with a JEOL 6063 L equipment and technique energy dispersive spectrometry (EDS) the distribution of alloying elements was evaluated in the biomedical component. The nanohardness with ultra-micro hardness tester Mitutoyo with maximum load of 100mN

Figure 1. Component of the Ti6Al4V alloy hot forged

Table I. Nominal chemical composition of the Ti6Al4V

Elements	(wt. %)
Al	5.5-6.5
V	3.5-4.5
Fe	0.25
C	0.08
O	0.13
N	0.05
H	0.012
Ti	Balance

RESULTS AND DISCUSSION

Figure 2 shows dark areas corresponding to α phase, while the clear zones correspond to particles of β phase dispersed in the array of α phase. Although the chemical attack employed did not reveal the contour of the grain α phase, we can infer that the grains are of type equiaxial from observation of the distribution of the particles β, which are located preferentially in these contours. This microstructure is primarily known as the type mill-annealed. In addition, it shows an optical micrograph of the longitudinal section of the bar where there is a lengthening of the grains in the direction of the forge, which coincides with the image in the horizontal axis. In the Figure 2 clear zones correspond to the grains of α and dark areas to the particles β.

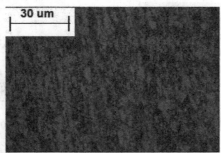

Figure 2. Microstructure of the forged alloy Ti-6Al-4V obtained by optical microscopy.

Figure 3 shown the microstructures in the majority of the sites indicate the type of structures stable α-β without defects of deformation, and some of shear bands. the material Ti-6Al-4V forging presents a type of equiaxed structure and an elongated shape unidirectional feature forged materials; in addition the bright areas are representative of the α phase, the phase darker is β. The alloying elements in this special material in the presence of vanadium in the alloy makes the biphasic (α + β) improving very significantly shaping by plastic deformation of the different types of prostheses.

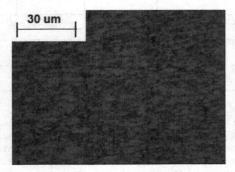

Figure 3. Micrograph obtained by optical microscopy of the Ti-6Al-4V alloy forged presenting a structure type "mill annealed".

On the other hand, this paper presents the test of nanoindentation to this material in different directions as is shown in Figure 4; in table 2 show the results the hardness and elastic modulus of the study material forged.

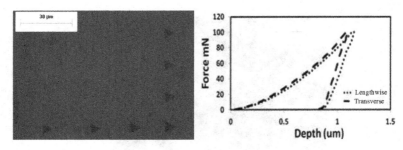

Figure 4. Nanoindentation test in different directions longitudinal and transversals to load 100 mN.

Table II. Properties of the material hot forged

	Modulus of Elasticity (GPa)		Hardness (Hv)	
	Longitudinal	Transversal	Longitudinal	Transversal
This Work	100	103	340±12.3	337.9±10.9
[14] Ti-6Al-4V	115		350.0	
[15] Ti-6Al-4V	101–110		346	
[16] Ti-6Al-4V	110		346	

The results obtained by X-ray diffraction are consistent with those found by optical microscopy in terms of the phases present in the material. In figure 5 are observed the areas of analysis on the material studied. In addition, shows figure 5 the family of levels of each of the phases which. The reflection of greater intensity for the α phase occurs at an angle of 40.42° corresponding to the (100) plane, while for the β phase occurs at an angle of 39.4° corresponding to the (110) plane, as shown in figure 5.

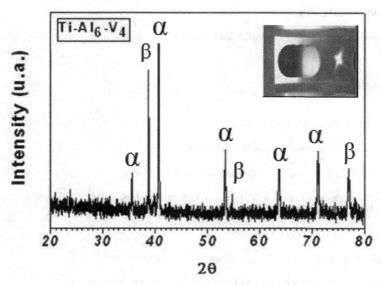

Figure 5. Areas of analysis by XRD an d XRD patterns

CONCLUSIONS

The effect of hot forging on the Ti-6Al-4V alloy shows the type of the morphology and microstructure, the analysis in this study by nanoindentation shows the hardness and elastic modulus of the phases present and XRD revealed type phases in the study. Furthermore shows a microstructure bi-modal or equiaxed structure optimal for this biomedical material. The module of elasticity and hardness were obtained at different directions with the type phase β and α presented a similitude in both directions. Also observed with the forging temperature a significantly influence on the microstructure and mechanical properties. The Diffraction patterns showing the presence β phase α. With the development of engineering forging parts or biomedical components using alternative manufacturing processes, has been shown to produce an improvement on the properties of Ti-6Al-4V alloy. This study shows the results of mechanical properties obtained of forging. These results also intended to contribute to the design and manufacture of biomedical prostheses.

ACKNOWLEDGMENTS

The author acknowledge PROMEP and COMECyT of Mexico for the support by this study.

REFERENCES

1. R. Raj., *Metall Trans,* **12**, 1089 (1981).

2. C. Chen, J.E. Coyne, *Metall Trans*, **7**, 1931 (1976).

3. J.A. Davidson, F.S. Georgette, *Technical Paper EM*, 87 (1986).

4. A. Astarita, A. Ducato, L. Fratini, V. Paradiso, F. Scherillo, A. Squillace, C. Testani, C. Velotti, *Engin Mater*, **557**, 359 (2013).

5. Y. Okazaki, *Mater,* **5**, 1439 (2012).

6. R.C. Picu, A. Majorell, *Mater Sci. and Engin.*, **326**, 306 (2002).

7. S. Bruschi, S. Poggio, F. Quadrini, M.E. Tata, *Mater Lett.*, **58**, 3622 (2004).

8. N.K. Park, J.T. Yeom, Y.S. Na, *Jour of Mater Process Techno*, **131**, 540 (2002).

9. G.G. Yapici, I. Karaman, Z.P. Luo, H. Rack, J. Scripta, *Mater,* **49**, 1021 (2003).

10. D. Eylon, J.A. Hall, C.M. Pierce, D.L. Ruckle, *Metall Trans.*, **7**, 1817 (1976).

11. N.K. Park, J.T. Yeom, Y.S. Na, *Jour of Mater Process Techno*, **131**, 540 (2002).

12. S.J. Li, T.C. Cui, Y.L. Hao, R. Yang, *Acta Biomater*, **4**, 305 (2008).

13. Y. Kim, E.P. Kim, Y.B. Song, S.H. Lee, Y.S. Kwon, *Jour of Alloy and Compo,* **603**, 207 (2014).

14. H.J. Rack, J.I. Qaz, *Mater Sci. and Engin.*, **26,** 1269 (2006).

15. M. Niinomi, *Mater Sci. and Engin.*, **243**, 231 (1998).

16. M. Niinomi, *Metall and Mater Trans*, **33**, 477 (2002).

Mater. Res. Soc. Symp. Proc. Vol. 1766 © 2015 Materials Research Society
DOI: 10.1557/opl.2015.407

Synthesis, Characterization and Antitumor Activity of 4-Ferrocenylpyridine-3, 5-Dicarbonitrile Derivatives and Sodium Polymeric Complexes Containing Carbanionic Ligands

E. Klimova[1], J. Sánchez[1], M. Flores[1], S. Cortez[1], T. Ramírez[2], A. Churakov[3], M. Martínez[2]

[1] Facultad de Química, UNAM, México D. F.,C.P.04500, México,

E- mail: eiklimova@yahoo.com.mx

[2] Instituto de Química, UNAM, México D. F.,C.P.04500, México.

[3] Institute of General and Inorganic Chemistry, Russian Academy of Sciences, Leninskii prosp., 31, Moscow 119991, Russia.

ABSTRACT

The reactions of 2-cyano-3-ferrocenylacrylonitrile with malononitrile in a ROH/H_2O medium in the presence of Na_2CO_3 afforded 6-alkoxy-2-amino-4-ferrocenylpyridine-3,5-dicarbonitriles, 6-alkoxy-2-amino-4-ferrocenyl-3-ferrocenyl-methyl-3,4-dihydropyridine-3,5-dicarbonitriles and Na^+ polymeric complexes: {[Na^+(2-ferrocenyl(tetracyano)propenyl)$^-$L]$_\infty$ and [Na^+(2-amino-3,5-dicyano-4-ferrocenyl-6-pyridyl-dicyanomethyl)$^-$L]$_\infty$ where L = ethanol, methanol. Complexes with L = acetonitrile (c), dimethylformamide (d), acetone (e), ethyl acetate (f) were prepared by recrystallization. The structures of the compounds 4b and Na^+ polymeric complexes 5c and 6d, e were established by the spectroscopic data and X-ray diffraction analysis. Two compounds 3a and 4a were tested *in vitro* against six human tumor cell lines U-251, PC-3, K-562, HCT-15, MCF-7 and SKLU-1 to assess their *in vitro* antitumor activity. The results suggest biological specificity towards PC-3, K-562 and HCT-15 cells for compound 3a, and towards PC-3 cell for compound 4a at doses 50 μM, which are lower than Cisplatin IC_{50}'s in the three cell lines.

INTRODUCTION

Recently, the interest in the chemistry of compounds of the ferrocene series has been increased due to their possible practical applications. Data on the industrial use of the ferrocene derivatives available from the patent literature [1, 2] reflect mainly their employment in military-industrial establishment and rocket technologies.

Publications demonstrating the use of ferrocene derivatives in nonlinear optics devices, chemistry of polymeric materials, synthetic organic chemistry (asymmetric synthesis) [3, 4], medicinal chemistry [5] supramolecular chemistry [6], chemo- and biosensors [7], materials science [8, 9], etc.

The ferrocene molecule itself is an energy rich structure. Therefore, studies on compounds comprising other energy rich fragments, such as ene, diene, and polyene chains, fused carbo- and heterocyclic systems, etc., in addition to the ferrocene moiety, are of special interest [10]. The incorporation of one or two iron-containing ferrocene substituents into a heterocyclic molecule will enlarge the spectrum of valuable characteristics. In particular, ferrocenyl-substituted pyridines have been extensively studied as ligands, in the synthesis of non-linear optical materials, etc. [11-13]. However, their biological activities have not hitherto been studied. Various methods to prepare ferroceno-containing pyridines have been reported [14]. The interest in pyridine compounds bearing ferrocenyl substituents in the molecules can be traced back to the discovery of ferrocene. In particular, biological activities of many nitrogen heterocycles, such as quinuclidines, pyrazolines, pyrazoles, pyrimidines, tetrahydropyridazines, bearing ferrocenyl substituents, have been reported [5]. It may be expected that ferrocenylpyridines and cyano(ferrocenyl)pyridines will also prove valuable, because they possess diverse biological activity, find use as potential bio-receptor ligands [15] new drugs [16] and significant intermediates for the synthesis of important materials [17]. For these reasons, development of new compounds containing cyano and ferrocenyl groups in the pyridines is strongly desired. In the present work, we report results from our investigations into reactions of the condensation of 2-cyano-3-ferrocenylacrylonitrile 1 with malononitrile 2 and of the tandem-transformations of 1 in alcohols/aqueous medium in the presence of bases and nucleophiles.

EXPERIMENTAL

All the solvents were dried according to standard procedures and were freshly distilled before use. Column chromatography was carried out on alumina (Brockmann activity III). The ^{1}H and ^{13}C NMR spectra were recorded on a Unity Inova Varian spectrometer (300 and 75 MHz) for solutions in CDCl$_3$ with Me$_4$Si as the internal standard. The IR spectra were measured with an FTIR spectrophotometer (Spectrum RXI Perkin - Elmer instruments) using KBr pellets. The mass spectra were obtained on a Varian MAT CH-6 instrument (EI MS, 70 eV). Elementary Analysensysteme LECO CHNS-900 was used for elemental analyses. The unit cell parameters and the X-ray diffraction intensities of 4b, 5c, 6d and 6e were recorded on a Gemini (detector Atlas CCD, Cryojet N$_2$) diffractometer. The structure of compounds 4b, 5c, 6d and 6e were solved by the direct method (SHELXS-97 [18]) and refined using full-matrix least-squares on F^2. 2-Cyano-3-ferrocenylacrylonitrile 1 was prepared by condensation of ferrocenecarbaldehyde with malononitrile in benzene in the presence of piperidinium acetate.

Reaction of 2-cyano-3-ferrocenylacrylonitrile 1 with malononitrile 2 in the presence of Na$_2$CO$_3$: Exp. 1: a mixture of compound 1 (1.13 g, 5.0 mmol), malononitrile 2 (0.4 g, 6.0 mmol), ethanol or methanol or (100 mL), H$_2$O (10 mL) and Na$_2$CO$_3$ (0.5 g, 5.0 mmol) was stirred and refluxed for 8 h. The solvents were removed *in vacuo* and the residue was dissolved in dichloromethane (50 mL). The solution was mixed with Al$_2$O$_3$ (activity III, 20 g) and the solvent was evaporated in air. This sorbent was applied onto a column with Al$_2$O$_3$ (the height of alumina is *ca.* 20 cm) and the reaction products were eluted from the column first with petroleum ether, then with a 2:1 hexane-dichloromethane to give compounds 3a,b, 4a,b, polymeric compounds 5a,b (with 6:1 dichloromethane – methanol) and 6a,b (eluent 1:2:1 dichloromethane - methanol – water) [19,20].

RESULTS AND DISCUSSION

All experiments were carried out using 2-cyano-3-ferrocenylacrylonitrile 1 with malono-nitrile 2 [~1:1.2 (exp. 1), ~1:3(exp. 2), respectively] in EtOH/H$_2$O or MeOH/H$_2$O medium in the presence of Na$_2$CO$_3$ by refluxed (Table 1). We found that following competitive processes occur upon this reaction: formation of 6-alkoxy-2-amino-4-ferrocenylpyridine-3,5-dicarbonitriles 3a,b multi-component condensation, cyclodimers 4a,b multi-component cyclodimerization, and formation of the Na$^+$ polymeric complexes: {[Na$^+$(2-ferrocenyl(tetracyano)propenyl)$^-$L]$_\infty$ 5a,b and [Na$^+$(2-amino-3,5-dicyano-4-ferrocenyl-6-pyridyl-dicyanomethyl)$^-$L]$_\infty$ 6a,b, where L = ethanol, methanol (Scheme 1).

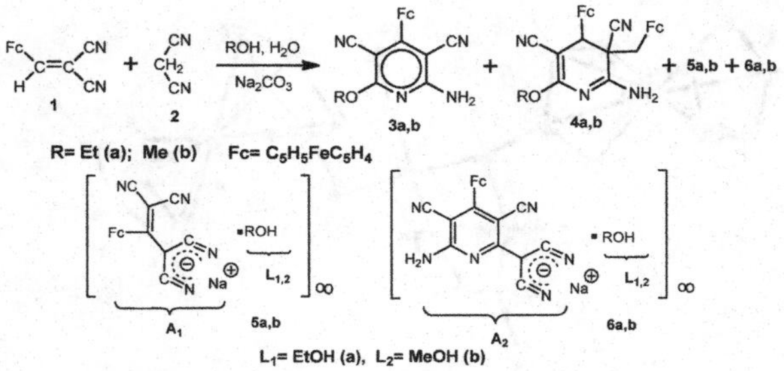

Scheme 1 Multicomponent competitive reactions of 2-cyano-3-ferrocenylacrylonitrile with malononitrile.

Table I. Reactions of 2-cyano-3-ferrocenylacrilonitrile 1 with malonitrile 2 in EtOH (a) or MeOH (b) in the presence of Na$_2$CO$_3$

Exp 1: FcCH=C(CN)$_2$ + CH$_2$(CN)$_2$				Exp 2: FcCH=C(CN)$_2$ + CH$_2$(CN)$_2$			
5 mmol		6 mmol		5 mmol		15 mmol	
EtOH ~ 70°C		MeOH ~ 60°C		EtOH ~ 70°C		MeOH ~ 60°C	
No	Yield (%)	No	Yield (%)	No	Yield (%)	No	Yield (%)
3a	48	3b	52	3a	6	3b	5
4a	19	4b	18	4a	6	4b	7
5a	7	5b	5	5a	21	5b	20
6a	6	6b	6	6a	45	6b	51

All products were isolated by column chromatography on alumina: compounds 3a,b and 4a,b with eluent 2:1 hexane - dichloromethane, 5a,b with eluent 6:1 dichloromethane - methanol and 6a,b with eluent 1:2:1 dichloromethane - methanol – water, and their structures were characterized by IR and NMR spectroscopy, mass spectrometry, and elemental analysis. The molecular structure of compound 4b was determined by X-ray diffraction analysis of their single crystals. The general view of molecule 4b is shown in Fig. 1. Key elements of the molecules 4b is the central six-membered ring with one nitrogen atom in the half-chair conformation. The N(1)-C(23) (for 4b) bond lengths is equal to $d = 1.319(3)$ Å. The ferrocenyl and ferrocenylmethyl substituents at C-4 and C-5 of 4b are *trans* oriented relative to the 6-membered cycle.

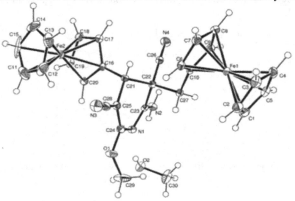

Figure 1 .The Crystal structure of 4b. CCDC- 878739.

Compounds 5a, b and 6a, b are red powders. Their structures were established by IR, ^{1}H and ^{13}C NMR spectroscopy and mass spectrometry. The IR spectra of compounds 5a, b and 6a, b contain bands at 2170-2225 cm[-1], which are characteristic of the cyano groups. The ^{1}H NMR spectra of these compounds contain characteristic signals for one ferrocenyl substituent and signals for one molecules of ethanol or methanol. In addition, the spectra of compounds 6a and 6b contains also signals for protons of the NH_2 groups. Data from ^{13}C NMR spectra of 5a, b and 6a, b corroborate the presence in each compound of one ferrocenyl fragment, one methyl group, four cyano substituents, and they also contain signals of one $C_{ipso}Fc$ carbon atoms, three and six quaternary carbon atoms, respectively. Mass-spectra of compounds 5a, b and 6a, b showed the signals corresponding to molecular ions [M]$^+$ (m/z = 325 and 391, respectively), [M+Na]$^+$ (m/z = 348 and 414, respectively), [M+Na+CH_3CH_2OH]$^+$ (m/z = 394 and 460, respectively) and [M+Na+CH_3OH]$^+$ (m/z = 389 and 446, respectively). On this basis, we supposed that compounds 5a,b and 6a,b may represent the sodium polymeric complexes containing carbanionic 2-ferroceny(tetracyano)propene and 3,5-dicyano-6-dicyanomethyl-(ferrocenyl)pyridine, respectively, and ethanol or methanol ligands (Scheme 1).Further, we observed that upon recrystallization of compounds 5a from CH_3CN, acetonitrile substitutes ethanol forming a new crystalline product 5c. Similarly, recrystallization of 6a from CH_3CN, $(CH_3)_2NCHO$, $(CH_3)_2CO$ and

CH$_3$COOC$_2$H$_5$ resulted in formation of compounds 6c, 6d, 6e and 6f, respectively (Scheme 2). Physicochemical characteristics of compounds 5c, 6c, 6d, 6e and 6f gave evidence that their structures differ from those of compounds 5a,b and 6a,b only by the presence of different ligands: CH$_3$CN, (CH$_3$)$_2$NCHO, (CH$_3$)$_2$CO or CH$_3$COOC$_2$H$_5$. Crystal structures of compound 5c, 6d and 6e were determined by X-ray diffraction analysis of single crystals. The structures 5c, 6d and 6e consist of substituted cyanoferrocenyl anions and sodium cations. In both compounds unsubstituted (Cp) and monosubstituted (Cp*) rings are planar within 0.006 Å.

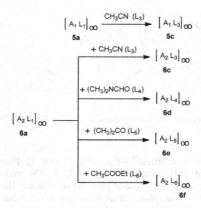

Scheme 2. Formation of complexes 5c and 6c-f

As expected, ferrocenyl fragments are approximately parallel. All Fe-C distances adopt ordinary values (Cambridge Structure Database, ver. 5.34 [21]).

Of interest, two malononitrile substituents in 5c are seriously unparallel. The dihedral angle between them is 36.6°. The only reason for this may be strong interaction of nitrile nitrogen atoms with Na cation, all cyano groups are linear with C-C≡N angles greater than 174.4(2)°. All four nitrile nitrogen atoms are involved in the interactions with Na$^+$ cation with the Na⋯N distances ranging within 2.4077(14)-2.5275(14) Å. The sodium cation has a slightly distorted octahedral coordination environment with *ortho*-angles varying within 82.23(5) °-97.55(5)°. Four of the six coordination sites are occupied by nitrile nitrogen atoms, while two others are engaged by solvent acetonitrile molecules. Two neighboring polyhedra are linked by bridging acetonitrile ligand forming center symmetric dimers (Fig. 2) with Na⋯ distances equal to 3.832 Å. In crystal, these dimers are combined by ferrocenyl anions in sophisticated 3D-packing.

To our best knowledge, anion 5c is the second example of bis(malononitrile)methanide anion in contrast to well-known to tris(malononitrile)methanide anions ⁻C[C(CN)$_2$]$_3$ (25 compounds in CSD).

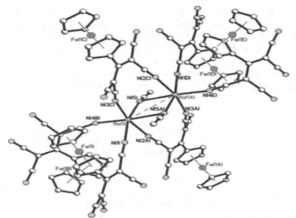

Figure 2. Bimetallic units in the structure of 5c. CCDC – 939093.

In 6d, Cp* ring is bonded to substituted 4-pyridyl cycle (Fig.3). The dihedral angle between the planes of pyridyl and cyclopentadyenyl rings is 33.0°. The amino group N (2) and cyano group C(19) lie approximately in the plane of pyridine substituent. However, cyano group C(20) is noticeably displaced from this plane. Thus, the angle between the center of pyridyl ring, atoms C (14) and C (20) is equal to 169.1°. Probably, this is caused by the strong interaction of N (4) atom with Na cation in the crystal lattice.

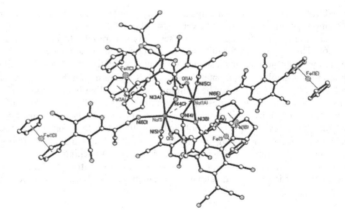

Figure 3 Bimetallic units in the structure of 6d. CCDC- 939092.

The central C (16) atom of the malononitrile moiety is planar (sum of valence angles is 359.6°) due to its negative charge and sp^2-hybridization. Malonitrile anion is sufficiently conjugated with pyridyl ring (torsion angle C (14)-C (15)-C (16)-C (17) is equal to -19.6°). All cyano groups in 6d are linear with C-C≡N angles greater than 174.5(3) °. All nitrile nitrogen atoms are involved in the interactions with Na cation (Fig. 3). However, nitrogen atoms N (1) and N (2) do not form coordination bonds with the cation. The N(1) atom is sterically hindered, while the N(2) is electronically depleted due to the strong conjugation with electronegative pyridyl cycle. The sodium cation has a distorted octahedral coordination environment with *ortho*-angles ranging within 78.52(7)-111.47(8)°. Five of the six coordination sites are occupied by nitrile nitrogen atoms with N⋯N distances vary within 2.419(2)-2.989(2) Å. The sixth position is engaged by solvent dimethylformamide molecule (N⋯O – 2.323(2) Å). Two adjacent sodium polyhedra share common edge forming dimeric center symmetric motif with Na⋯Na separation equal to 3.724 Å (Figs. 3). In crystal, these dimeric units are linked by ferrocenyl anions in complicated 3D-network.

Pharmacology

In order to examine the applicability of four compounds (3a-6a) as antitumor agents, they were tested *in vitro* against six human tumor cell lines: U-251(glioma), PC-3 (prostate), K-562 (leukemia), HCT-15 (colon), MCF-7 (breast) and SKLU-1 (lung) [22]. A primary screening at a fixed concentration showed cytotoxicity against the six human tumor cell lines tested, besides human normal lymphocytes (MT2) and macrophages murinos (MTT). Cisplatin was used at the same concentration as a positive control, were also determined at 50 μM in DMSO (Table II).

Table II. Inhibition of human tumor cells lines by compounds 3a and 4a, besides human lymphocytes (MT2) and macrophages murinos (MTT) at 50 μM in DMSO[a] and by compounds 5a and 6a at 50 μM in H₂O

No	3a	% of growth 4a	inhibition 5a · (H₂O)	6a · (H₂O)	Cisplatin
U-251	77.8±14	84.6±10.	9.6±1.5	22.2±1.2	89.9±8.1
PC-3	93.4±3.3	96.3±3.1	34.3±4.2	20.2±5.1	86.7±4.1
K562	96.1±3.8	86.3±14	35.3±11.	9.8±1.6	74.4±2.1
HCT-15	91.9±3.0	75.5±15	6.7±1.3	8.0±2.2	81.8±7.1
MCF-7	66.3±9.2	57.2±9.4	12.0±9.7	10.3±5.0	77.9±2.5
SKLU-1	88.2±7.2	64.9±8.1	21.3±11	12.2±6.8	95.8±2.1
MT2	88.72±4.1	94.58±5.1	34.68±14.2	37.95±5.1	70.07±0.4
MTT	93.90±6.0	87.03±7.5	82.58±8.1	75.75±4.3	72.01±3.3

Results express mean ± standard error (SEM) obtained from 2 independent experiments performed at 48 h.

Compound 3a showed a 93.4, 96.1 and 91.9 percent inhibition of cellular growth at 50 μM for three human tumor cell lines (PC-3, K562 and HCT-15, respectively) and also 88.72 percent inhibition of besides human lymphocytes (MT2), compound 4a showed better activity than cisplatin for PC-3, K562 and MT2 (Table II).

CONCLUSION

Reactions of 2-cyano-3-ferrocenylacrylonitrile with malononitrile (~1:3) in a $EtOH/H_2O$ or $MeOH/H_2O$ medium in the presence of Na_2CO_3 gave novel 6-alkoxy-2-amino-4-ferrocenylpyridine-3,5-dicarbonitriles 3a,b (products of multi-component condensation), 6-alkoxy-2-amino-4-ferrocenyl-3-ferrocenylmethyl-3,4-dihydropyri-dine-3,5-dicarbonitriles 4a,b (products of multi-component cyclodimerization) and polymeric complexes of sodium with organometallic moieties of the general formulas $[Na^+(2\text{-ferrocenyl-(tetracyano)propenyl})^-L]_\infty$ 5a-c and $[Na^+(2\text{-amino-3,5-dicyano-4-ferrocenyl-6-pyridyl-dicyanomethyl})^-L]_\infty$ 6a-f, where L = ethanol (a), methanol (b), acetonitrile (c), dimethylformamide (d), acetone (e), and ethyl acetate (f). These complexes were obtained in high yields. The new complexes were structurally characterized by elemental analysis, IR, ^1H and ^{13}C NMR spectroscopy, Mass-spectrometry, and single crystal X-ray diffraction analysis (supplementary data). The synthesized compounds 3a-6a was evaluated for their *in vitro* anticancer activities against six human tumor cell lines: U-251, PC-3, K-562, HCT-15, MCF-7, SKLU-1 and MT2. Compounds 3a and 4a exhibited the highest activity against three tumoral cell lines y cisplatin, which was used as reference.

ACKNOWLEDGEMENTS

This work was supported by the DGAPA – UNAM (Mexico, grant IN 211112). Thanks are due to Eduardo Arturo Vázquez López for his technical assistance.

SUPPLEMENTAL MATERIAL

Supplemental data for this article can be accessed on the publisher's website.

REFERENCES

1. E.G. Perevalova, M.D. Reshetova, K.I. Grandberg, Methods of Organometallic Chemistry: Organoiron Compounds, *Ferrocene,* Nauka: Moscow, 437, (1983).
2. A. Togni, T. Hayashi, Ferrocenes, VCH, Weinheim, 433, (1995).
3. G. Salemand, C.L. Raston, The Use of Organometallic Compounds in Organic Synthesis, F. R. Hartley (Ed). Wiley: Chicheester. 159, (1987).
4. M.E. Kuehne, V.K. Bandarage, *Journal Organic Chemistry,* **61**, 1175 (1996).
5. N. Metzler-Nolte, *Angewandte Chemise International,* **40**, 1040 (2001).
6. A.E. Kaifer, J.C. De Mendoza, Comprehensive Supramolecular Chemistry, Elsevier, Oxford **1**, 701 (1996).
7. T. Yao, G.A. Rechnitz, *Biosensors,* **5**, 307 (1987).
8. H. Zhu, H. Lin, H. Guo, L. Yu, *Materials Science and Enginnering,* **138**, 101 (2007).
9. P. Stepnicka, Ferrocenes: *materials and biomolecules,* **1**, 499 (2008).
10. G.M. Martinez, B.T. Klimova, E.I. Klimova, A. Martin, Cato (ed), Leading Edge Organometallic Chemistry Research, USA 27, (2006).
11. T.M. Miller, K.J. Ahmed, M.S. Wrighton, *Inorganic Chemistry,* **28**, 2347 (1989).
12. J. Rajput, A.T. Hutton, J.R. Moss, H. Su, Ch. Imrie, *Journal Organometallic Chemistry,* **691**, 4573 (2006).

13. I.P. Beletskaya, A.V. Tsvetkov, G.V. Latyshev, V.A. Tafeenko VA, Lukashev, *Journal Organometallic Chemistry*, 637-639, 653-663, (2001).
14. M.G.A. Schvekhgeimer, *Russian Chemical Reviews,* **65**, 66 (1996).
15. C. Biot, N. Chavain, F. Dubar, B. Pradines, X. Trivelli, J. Brocard, I. Forfar, D. Dive, *Journal Organometallic Chemistry,* **694**, 845 (2009).
16. C.A. Brijnincx P, P.J. Sadler, Current Opinion in Chemical Biology, **12**, 197 (2008).
17. K. Kowalski, R.F. Winter, *Journal Organometallic Chemistry,* **700**, 58 (2012).
18. G.M. Sheldrick, SHELXS-97, Program for the Refinement of Crystal Structures, University of Göttingen, Germany, (1994).
19. E.I. Klimova, M. Flores Álamo, S. Cortéz Maya, M. Martínez Klimova, L. Ortiz Frade, T. Klimova, *Molecules,* **17**, 10079 (2012).
20. E.I. Klimova, M. Martínez García, M. Flores Álamo, A.V. Churakov, S. Cortéz Maya, I.P. Beletskaya, *Polyhedron,* **68**, 272 (2014).
21. F.H. Allen, The Cambridge structural Database, **58**, 380 (2002).
22. A. Monks, D. Scudiero, P. Skehan, R. Shoemaker, K. Paul, D. Vistica, C. Hose, J. Langle, P. Cronise, A. Vaigro-Wolff, M. Gray Goodrich, H. Campbell, J. Mayo, M. Boyd, *Journal of the National Cancer Institute,* **38**, 757 (1991).

Mater. Res. Soc. Symp. Proc. Vol. 1766 © 2015 Materials Research Society
DOI: 10.1557/opl.2015.408

Study of Microstructure and Mechanical Properties of an Ankle Prosthesis Removing

J. G. Flores Becerra[1], N. López Perrusquia[1*], M. A. Doñu Ruiz[1], A. López Perrusquia[3,] J.V.
Cortes Suarez[2].

[1]Universidad Politécnica Valle de México; *Grupo Ciencia e Ingeniería de Materiales*, UPVM,
Tultitlán. Edo de México
*E-mail: noeperrusquia@hotmail.com
[2]Universidad Autónoma Metropolitana Azcapotzalco Avenida San Pablo 180, Azcapotzalco,
Reynosa Tamaulipas, 02200 Ciudad de México, Distrito Federal
[3]Instituto Nacional de Rehabilitación (INR), Calz. México Xochimilco No. 289 Col. Arenal de
Guadalupe, CP14389, México

ABSTRACT

This work studies the change microstructural and mechanical properties of an ankle
prosthetic material 316LVM stainless steel, retired from a 36 year old patient. The medical grade
316LVM stainless steel was characterized by scanning electron microscopy (SEM), optical
microscopy (OM), X-ray diffraction (XRD), hardness Rockwell C (HRC) and nanoindentation
tests. The results showed that the ankle prosthesis has different microstructural change along the
implant and presence of corrosion pits with inclusions, the mechanical properties like modulus
elasticity and hardness decrease.

INTRODUCTION

One of the most prominent areas of biomaterials is their application orthopedic and
implants [1]. Biomedical metals and alloys are used to replace certain parts of the human body,
in order to replace the affected area, in addition to regenerate their mobility to 100 percent of the
patient [2]. Thus it can be said that the metallic materials are essential, at present, also for those
clinical applications in orthopedics, which require support static or dynamic load with
appropriate microstructural characteristics, appropriate mechanical properties and increased
corrosion resistance the human organism [3]. However the number of metal alloys that can
withstand changes in the human body and the human body are very low [4]. One of the important
parts of the microstructural study of biomedical and orthopedic materials, is the change in the
human organism, such as anesthesia drugs, antiseptics and antibiotics to name a few, the release
of chemical species to the human body o degenerative rheumatic diseases; addition of different
diseases affecting the microstructure of the material, triggered a loss in mechanical properties
and causing the withdrawal of the biomedical component of complications by the microstructural
change [5]. Also for its change in mechanical properties; this gives as a result a change in the
patient's mobility joint pain and eventually immobility of the affected limb [6]. The materials
stainless steel, cobalt-chromium alloy and pure titanium or should have a microstructural
features that support the changes of the human body and so aggressive media is the human body
to the surrounding tissues [7]. The microstructure plays an important role as this tend to improve
their behavior in the human body, which make the optimum microstructure allow increased

component life, well that is bioactive metal substrate, which allows its union with the surrounding tissues [8]. Therefore, this study evaluates the microstructural change of a biomedical component ankle where the microstructural variation is observed, in order to determine the possible cause of implant removal

EXPERIMENTAL

The research has developed a component essentially removed from ankle of 36 years old patient, the implant was withdrawn by discomfort after two years, medical grade 316LVM stainless steel ankle prosthesis, provided by the Institute National of Rehabilitation (INR), extracted in the operating room area national institute of rehabilitation, the patient requested implant removal due to pain in his mobility. Microstructural characterization was determined by means of a metallographic microscope Olympus GX 51 along the ankle prosthesis, also by scanning electron microscopy with a JEOL 6063 L equipment and technique energy dispersive spectrometry (EDS) the distribution of alloying elements was evaluated in the biomedical component. The Rockwell C hardness was used tester Mitutoyo and instrument nanohardness with ultra-micro hardness tester Mitutoyo with maximum load of 150mN.

RESULTS AND DISCUSSION

The implant grade stainless steel extracted from biomedical female patient, was observed in situ general degradation, confirming the degradation by cellular body fluids due to the presence of organic compounds and inorganic, as can be observed in Figure 1 the photographs taken after the removal of the ankle prosthesis, showing the presence of surfaces with corrosion process.

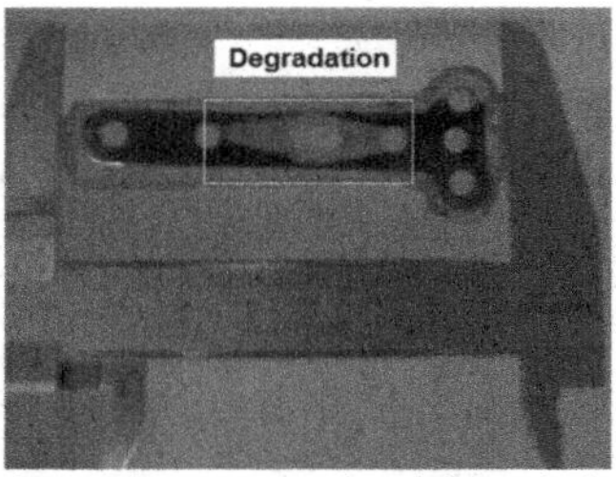

Figure 1. Implant removed and degradation zone.

Once extracted and evaluated the prosthesis in situ, we proceeded to microstructural characterization, in Figure 2, it shown the microstructure along the ankle prosthesis.

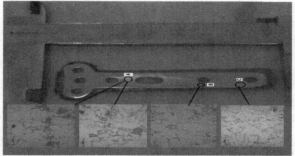

Figure 2. Microstructures of retired biomedical component.

In Figure 2 shown micrographs in different zone corresponding implant where the implant deterioration degree shown in all areas, showing that there is a further degradation and corrosion potential well underway so the patient has a diagnosis fever and inflammation where the prosthesis, possibly a downgrade compatibility of bodily fluids from your body with this implant was because the material was removed at the request of patient discomfort and the diagnostic by personnel of INR.

Figure 3 shown microstructure along the ankle prosthesis on the degradation zone, it was observed matrix a fully austenitic with grains equiaxed, typical grains in this steel and the presence of deformation twins also possible corrosion was observed in the material due a body fluids.

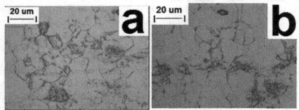

Figure 3. Microstructures of 316 LVM stainless steel along the specimen.

In figure 4 the SEM micrograph of the cross section of the sample taken is shown, where the austenitic grain is identified, and a corrosion potential of the material, it also presents a possible corrosion in grain as it shown in figure 4a. The corrosion possibly reacts with the patient, resulting in inflammation in the member where the implant and fever, which led to the extraction of biomedical implant. In the figure 4b shown the effect of corrosion in different areas the implant removed the patient.

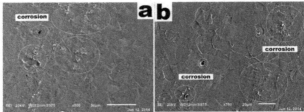

Figure 4. Micrographs (a) center and (b) left, of the areas affected of removed implant.

Figure 5a described the deterioration of the material presented AISI 316 LVM stainless steels on degradation zone. Also shown the affected by corrosion of biomedical implant. In figure 5b is presented the EDS analysis on the part of degradation material, where he observed the distribution of alloying elements and possible inclusions that are derived a deteriorate on material.

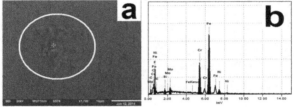

Figure 5. (a) Micrograph obtained by SEM (b) EDS distribution of elements in the area affected by corrosion.

Figure 6 shows tests to study material nanohardness. The nanohardness obtained along the specimen show a significant change in the elastic modulus and hardness as shown in Table 1. Also show decrease of the mechanical properties of the implant removed. The change of mechanical properties obtained by nanoindentation and indentation HRC, it is assumed by degradation of the material by the human body fluid, this shows the poor compatibility of the fluids of the patient with the material implanted in his body. However shown comparative in table 1 with different mechanical properties studies by Aperador *et* al. [9] and Multigner *et* al. [10-11].

Figure 6. Nanoindentation test with 150 mN load in the implant removed.

The behavior of the hardness and elastic modulus of the removed component, compared with other studies are showed in Table 1.

Table 1 Mechanical properties of 316LVM stainless steel

Material	Modulus Elastic (GPa)	Hardness (HRC)	Hardness (Hv)
W. Aperador et al. [9]	203	30	307
A. Krawczynska et al. [12]	200-205	43	421
R. Narayan et al. [13]	192-201	23.1-42.9	257.9-419.8
Present work	186.2	6.73	181

Figure 7 shown the results of X-ray diffraction, and shown peaks of the corrosion products on the surface of this material such as chromium oxide (Cr_2O_3) and iron oxide (Fe_2O_3) since the stainless steel has two important features from the viewpoint of corrosion resistance and passivation capacity.

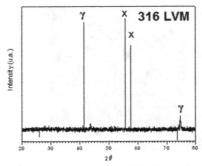

Figure 7. XRD plot showing presence of austenite 316 LVM stainless steel (γ-austenite) and as oxides (X- Cr_2O_3).

CONCLUSIONS

It was shown that modification of retired biomedical component on the surface of stainless steel 316LVM, a significant change was found in the microstructure presented along the ankle prosthesis, also the type and morphology shown in micrographs are derived from the human body fluids and active material removed on body tissues. The stainless steel is a material with high resistance to degradation, such corrosion, however presents a low resistance to corrosion because the implant was removed ahead of schedule of the patient. Likewise it has a deteriorated toughness of the material having a decrease over the entire implant and property loss, this loss of strength and material properties has a severe disease in the implant as shown in this study. The study shows degradation and corrosion the 316 LVM, by scanning electron microscopy, optical microscopy and mechanical tests. The EDS shown distribution of the alloying elements in the areas of degradation; will have the chromium and the iron oxide forming and the corrosion of the material present.

ACKNOWLEDGMENTS

The author acknowledge PROMEP and COMECyT of Mexico for the support by this study.

REFERENCES

1. Van Noort R, *Journal of Materials Science* 22: 3801-381. (1987).

2. Fraker A. "Corrosion of metallic implants and prosthetic devices", ASM International Metals Handbook, 4th ed.pp. 1324. (1992).

3. Zhang Zhi-Qiang, Dong Li-Min, Yang Yang, Guan Shao-Xuan, Liu Yu-Yin, Yang Rui, *Trans Nonferrous Met Soc* 22: 2604−2608. (2012).

4. Bou-Saleh Z., Shahryari A., Omanovic S. *Thin Solid Films* 515: 4727-4737. (2007).

5. Thamaraiselvi T. V. and Rajeswari S. *Trends Biomater Artif Organs* 18: 242-246. (2005).

6. Marija Mihailović, Aleksandra Pataric, Zvonko Gulišija, Djordje Veljović, Djordje Janaćković. *Chem Indu & Chem Engin Quarter* 1: 45−52. (2011).

7. Prakash Sojitra, Chhaya Engineer, Devesh Kothwala, Ankur Raval, Haresh Kotadia, Girish Mehta. *Trends Biomater Artif Organs* 23:115-121. (2010).

8. Jan Sieniawski, Ryszard Filip, Waldemar Ziaja Materials & Design,18: 361-363.(1997).

9. Aperador W., Melgarejo M., Ramírez–Martin C., The *man and the machine,* 38:51-58. (2012).

10. Multigner M., Frutos E, González-Carrasco J. L, Jiménez, J. A. Marín P, Ibáñez J., *Materials Science and Engineering* 29:1357–1360. (2009).

11. Multigner M., Ferreira-Barragáns S, Frutos E., Jaafar M., Ibáñez J., Marín P., Pérez-Prado M. T., González-Doncel G., Asenjo A., González-Carrasco J. L., *Surface & Coatings Technology* 205:1830–1837. (2010)

12. Krawczynska A. T, Brynka T, Gierlotka S, Grzanka E., Stelmakh S, Palosz B, Lewandowskaa M, Kurzydlowski. K J. *Mechanics of Materials* 67:25–32. (2013)

13. Narayan R. ASM Handbook, Materials for Medical Devices, Medical Implant Materials 23: 200-210. (2012)

Characterization of Welded Materials

Mater. Res. Soc. Symp. Proc. Vol. 1766 © 2015 Materials Research Society
DOI: 10.1557/opl.2015.409

Microstructural Effects between AHSS Dissimilar Joints Using MIG and TIG Welding Process

G.Y. Pérez Medina[1], M. Padovani[2], M. Merlin[2], A.F. Miranda Pérez[1], F.A. Reyes Valdés[1]

[1] Corporación Mexicana de Investigación en Materiales (COMIMSA)
E-mail: gladysperez@comimsa.com
[2] Engineering Department, University of Ferrara, Via Saragat 1, 44122, Ferrara, Italy

ABSTRACT

Gas tungsten arc welding-tungsten inert gas (GTAW-TIG) is focused in literature as an alternative choice for joining high strength low alloy steels; this study is performed to compare the differences between gas metal arc welding-metal inert gas (GMAW-MIG) and GTAW welding processes. The aim of this study is to characterize microstructure of dissimilar transformation induced plasticity steels (TRIP) and martensitic welded joints by GMAW and GTAW welding processes. It was found that GMAW process lead to relatively high hardness in the HAZ of TRIP steel, indicating that the resultant microstructure was martensite. In the fusion zone (FZ), a mixture of phases consisting of bainite, ferrite and small areas of martensite were present. Similar phase's mixtures were found in FZ of GTAW process. The presence of these mixtures of phases did not result in mechanical degradation when the GTAW samples were tested in lap shear tensile testing as the fracture occurred in the heat affected zone. In order to achieve light weight these result are benefits which is applied an autogenous process, where it was shown that without additional weight the out coming welding resulted in a high quality bead with homogeneous mechanical properties and a ductile morphology on the fracture surface. Scanning electron microscopy (SEM) was employed to obtain information about the specimens that provided evidence of ductile morphology.

INTRODUCTION

The automotive industry is constantly under several challenges in many aspects, such as develop of new materials and improve their manufacturability. In order to achieve light weight, reduced emissions and ensure conductor safety, advanced high strength steels (AHSS) are able to fulfill these requirements. Transformation-induced plasticity and martensitic steels are well suited for light weighing car body constructions, even for advantage to reduce the safety problems. The second challenge is welding these materials, it is necessary invest sources on welding areas for some conventional and advanced technologies. In recent years these steels have been subject of extensive interest because of their unique combination of mechanical properties, combine high strength with good formability.

Their microstructure consist bainite (B), ferrite (F), martensite (M), and retained austenite (RA) [1-3]. Beside martensitic steels (MS) with strength level ≥ 1000 MPa, contains different combinations of C, Mn, Si, Cr, Mo, V, Ti, Nb, Cu, Ni, etc., to obtain adequate hardenability and high strain hardening capacity, their microstructure consist of martensite matrix. The continuous necessity to lighten the automotive body turns out to be priority due to the rising cost of petroleum. At least 1% reduction of body weight may result in a 0.5% decrease of fuel consumption [4]. Therefore, it is only comprehensible to utilize high resistant materials in order to reduce the cross sectional area (thickness), hence to diminish the structural weight.

Nevertheless, the welding processes, due to their thermal cycle, might inflict alterations or even damages on the tensile properties of structural material [5-7]. The use of relatively thin TRIP and MS steels sheets in the manufacture of car bodies requires of homogeneous mechanical properties in addition to a relatively high corrosion resistance. Hence, a reduction in thickness and/or microstructural changes due to welding effects can be detrimental for the mechanical integrity of these steels. Thus, it becomes necessary to establish the effect of welding processes and parameters on the resultant microstructures and on the exhibited mechanical properties. Furthermore this investigation assesses the application of GTAW to advanced high strength steels and to compare results with those obtained by GMAW processes.

EXPERIMENTAL

Table I gives the chemical composition of the TRIP 780 steel and MS steel. Table 2 gives the exhibited mechanical properties; the investigated steels were supplied in the form of 2.7 mm and 1.5 mm steels thick sheets respectively. The dissimilar joints were carrying out using manual gas tungsten arc welding without any filler metal, obtaining autogenous welds. Table 3 describes the welding parameters applied for GTAW and GMAW processes. Welding was carried out using a Miller Syncrowave 250 AC/DC in COMIMSA training center. Overlap length was established to 20 mm in order to obtain tensile specimens similar to standard sheet-type 12.5 mm wide contained in ASTM E8/E8M 09. Tensile tests were carried on using a Tinius-Olsen SuperL 120/602 with maximum force of 600 kN. In the case of GMAW process was carried out using a 6 axes KUKA KR-16 robot, connected to a Lincoln Electric Powerwave 455 MIG/MAG welder through a PW-455X analogical interface, the filler metal was of the ER100S –G type (Table 1-2) with a 0.9 mm diameter suitable for high tensile strength steel (980 MPa), according to the classification contained in AWS 5.28/5.28M:2005 "Specification for low-alloy steel electrodes and rods for gas shielded arc welding". A stick-out length of 9 mm was used; a torch travel angle of 15 deg was kept in order to containing heat input and obtaining a deeper penetration. The arc was shielded by a 99.99% Argon gas with a flow rate of 50 cubic feet per hour (CFH, 1 m^3/h).

Specimens for microstructural analysis were taken by cold cutting the sheets. They were embedded in epoxy resin, hand grounded and polished. Silicon carbide abrasive grid was used for grinding following this sequence: 80-320-800-1200. After grinding, samples were hand polished with 1 μm diamond paste and using ethanol as lubricant. The etching solutions used was Klemm´s I (1 g $K_2S_2O_5$ potassium metabisulfite, 79.4 g $Na_2S_2O_3$-5H_2O Sodium Thiosulfate) and 2 % Nital followed by heat tinting. OLYMPUS microscope with CCD camera INFINITY2 and INFINITY ANALYZE software were adopted for capturing microstructure. Furthermore, the specimens were analyzed at the scanning electron microscope (SEM).

Table I. Chemical composition of the filler metal and AHSS steels.

Wt%	C	S	Mn	P	Si	Cr	Ni	Mo	Cu	V	Nb	Ti
TRIP 780	0.1	<0.002	1.98	<0.002	2.35	<0.06	<0.04	<0.03	0.012	0.019	<0.004	0.01
MS	0.08	0.005	0.54	0.009	0.02	0.01	0.04	0.02	0.012	0.004	---	---
ER100S-G	0.06	0.005	1.96	0.009	0.50	0.06	1.96	0.45	0.14	0.01	---	0.04

Table II. Mechanical properties of the filler metal and AHSS steels.

Base Metal	Yield Strength (MPa)	Ultimate Tensile Strength (MPa)	Elongation (%)	Hardness (Hv)
TRIP 780	652	854	20	254
MS	917	1030	10	310
ER100S-G	730	980	20	350

Table III. Parameters applied.

ID sample	Current (A)	Voltage (V)	Time (s)	Travel speed (mm/min)	Heat input (J/mm)	Force max (kN)
GTAW	60	6	20	76.2	141.7	10.25
GMAW	172.4	20	15	500	331.0	3.25

RESULTS AND DISCUSSION

Microstructure

Figure 1 shows the microstructural features of the TRIP and martensitic steels in the as received condition. Notice the mixture of ferrite, bainite and martensite phases in both steels including retained austenite particularity in TRIP steel [8, 9]. The quantitative determinations of the various phases contents in the base metal (BM), fusion zone (FZ) and heat affected zone (HAZ) are given in Figure 2 for every process, including the optical micrographs. From these results, it can be observed that the percent of M in the FZ goes up to 10 % in GTAW process compare with 2 % in the FZ of GMAW process. If the steel is quenched rapidly enough from the austenitic field there is insufficient time for eutectoidal diffusion-controlled decomposition processes to occur, and the steel transform to martensite or in some cases martensite with a few per cent of retained austenite. In the case of GTAW process, the cooling rate is such that the majority of carbon atoms in solution in the fcc γ-Fe remain in solution in the α-Fe phase. Martensite is thus simply a supersatured solid solution of carbon in α-Fe [10]. In contrast, the amount of bainite goes up 94% in the FZ of GMAW process, the bainite plates form by a shear mechanism in the same way as the growth of martensite plates. At sufficiently low temperatures the microstructure of bainite changes from laths into plates and the carbide dispersion becomes much finer, rather like in tempered martensite. There are not significant differences of amount phases in the HAZ for both base metals.

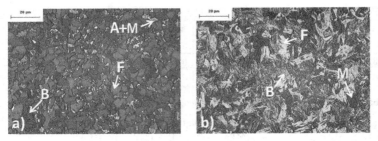

Figure 1. Optical micrograph of the AHSS steels BM: a) TRIP780 and b) MS. Etchant: Klemm`s and natal

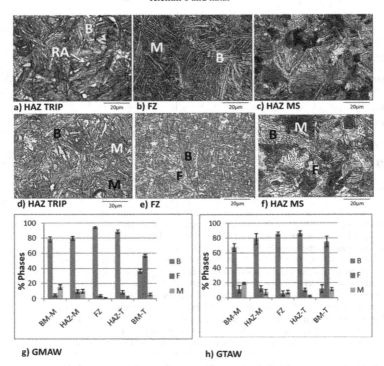

Figure 2. Optical micrographs for GTAW process: a), b) and c); optical micrographs for GMAW process: d), e) and f); percent phases in the two processes: g), h). Etching applied Klemm´s I solution.

Microhardness and Lap shear tensile testing

Dissimilar configuration has given a good response to weldability because of the excellent melting rate of the martensitic top sheet, especially when using high arc currents. Apparent ductile fracture morphology was found on specimens. Cracks clearly occurred between FZ-HAZ interfaces or within the heat affected zone. It can be affirmed that the weld zone was stronger than the heat affected zone [11, 12]. The thermal cycle causes changes in properties of the base material in the heat-affected zone due to the combination of phase changes and thermal/mechanical stress. Furthermore, the specimens that provided evidence of ductile morphology were analyzed at the scanning electron microscope. As can be seen from Figure 3 the samples exhibited a fracture surface with a great amount of dimples, which are oriented in different directions because of the asymmetrical straining. In particular larger voids are aligned along the crack propagation, or the direction given by the bending forces in the top plate. Microhardness profiles of the welded regions indicated that the hardness in HAZ of TRIP steel was relatively elevated in the GMAW process specimen, near 400 Hv compare with 355 Hv in the HAZ of TRIP steel GTAW process (Figure 4), indicating the development of martensite in this region, equiaxed and fine grain microstructure created in the HAZ by the influence of chemical composition of filler metal using the GMAW process. Although the heat input in GTAW process is almost reduce twice the heat input of the GMAW process (Table 3), this does not influence in the development of martensitic structure in the HAZ of TRIP steel welded by GTAW process [13].

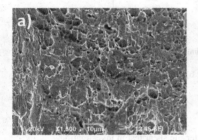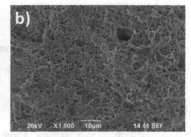

Figure 3. SEM micrographs of fracture surfaces: a) GTAW, b) GMAW.

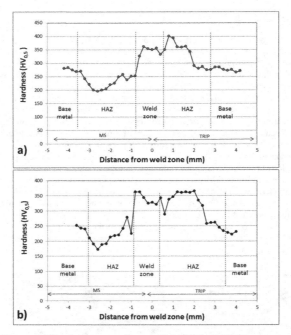

Figure 4. Microhardness profiles: a) GMAW, b) GTAW processes.

CONCLUSIONS

Microhardness measurements were combined with tensile shear test in investigating the weldability of dissimilar TRIP to martensitic steels sheets using GTAW and GMAW processes. It was found that welding using GTAW process leads to development of relatively high contents of martensite (> 9 %). The highest amounts of martensite were found in the FZ, whereas GMAW process promoted the development of a mixture of predominantly bainite and ferrite phases in FZ. In both welding processes, no embrittlement could be found in the FZ regions which can be attributed to the formation of martensite. On the other hand, the shear tensile test indicated that using GMAW process with a filler metal ER100S-G with a ultimate tensile strength of 980 MPa, value between TRIP and MS steels UTS, provided of low mechanical properties in comparison with the properties obtained by GTAW, which is an autogenous process, where it was shown that without additional weight the out coming welding resulted in a high quality bead with homogeneous mechanical properties and a ductile morphology on the fracture surface. In the GMAW process, the microhardnes profiles did not follow the same trends as in GTAW process. In this case, maximum Vickers hardness values of up to 400 HV where found in the HAZ of TRIP steel. In particular, it is found that the metal in the HAZ regions of the GMAW process, exhibited a significant increase in hardness.

REFERENCES

1. L. Samek, E. De Moor, J. Penning, J.G. Speer, B.C. De Cooman, *Metallurgical and Materials Transactions A*, **39,** 2542 (2008).
2. K. De Amar, J.G. Speer, D.K. Matlock, *Advanced Materials and Processes,* **161,** 27 (2003).
3. M. Zhang, F. Zhu, D. Zheng, *Journal of Iron and Steel Research, International*, **18,** 73 (2011).
4. M. Maurizio Mini, *Expert System with Applications,* **38,** 7911 (2011).
5. M.D. Tumuluru, *Welding Journal*, **85,** 31 (2006).
6. S. Oliver, T.B. Jones T.B, Fourlaris G, *Mater Charact.* , **58,** 390 (2007).
7. S. Dinda, R. Diaz, *Advanced Technologies and Processes 5 (1995).*
8. G. Thewlis, *Materials Science and Technology*, **20,** 143 (2001).
9. J.C. Villavicencio, *Tésis Relación, microestructura/propiedad en la soldadura GTAW entre aceros inoxidables y aceros al carbono*, (2010).
10. D.A. Porter and K.E. Easterling, *Taylor and Francis*, 383.
11.T. Mohandas, G. Madhusudan Reddy, B. Satish Kumar, *Journal of Materials Processing Technology*, **88,** 284 (1999).
12. S. Yaowu, H. Zhunxiang, *Journal of Materials Processing Technology,* **207,** 30 (2008).
13. M. Zhang, L. Li, R. Fu, J. Zhang, Z. Wan, *Journal of Iron and Steel Research, International,* **15,** 61 (2008).

Mater. Res. Soc. Symp. Proc. Vol. 1766 © 2015 Materials Research Society
DOI: 10.1557/opl.2015.410

Effect of Gas Metal Arc Welding (GMAW) Parameters on Wear Behavior of Heat Affected Zone of HSLA Steel Plates

Z.L. López Bustos[1], F.J. García Vázquez[1], G.Y. Pérez Medina[1], B. Vargas Arista[2] and V.H. López Cortez[1]

[1] Corporación Mexicana de Investigación en Materiales (COMIMSA), Calle ciencia y tecnología No. 790, Col. Saltillo 400, cp. 25290, Coahuila, México

E-mail: zayra.lopez@comimsa.com

[2] Instituto Tecnológico de Tlalnepantla, División de Estudios de Posgrado e Investigación, Av. Instituto Tecnológico s/n, Col. La Co-munidad, Tlalnepantla de Baz, Edo. de México, México 54070.

ABSTRACT

The wear phenomenon may occur for a variety of work conditions in the material. It causes losses in terms of time and costs in the components which are used for heavy machinery due to its re-pair or even replacement. It is important to select suitable materials that exhibit high-quality weldability and resistance to abrasive wear such as the high strength low alloy (HSLA) steel grade 950A. Therefore, it is necessary to study the wear behavior of this kind of steel after components are joined by multi-pass gas metal arc welding (GMAW) process, specifically on the heat affected zone (HAZ). The aim of this research was to evaluate wear resistance by pin on disc test and hardness on heat affected zone of HSLA steel plates with thickness of 14 mm joined by using GMAW process varying different parameters as wire feed speed and voltage. The influence of microstructural features such as carbide precipitation on wear behavior and hardness was investigated using optical microscopy (OM) and scanning electron microscopy (SEM). The results show that microstructure is modified by the heat input of the welding process, affecting the material properties and causing more susceptibility to wear on the welded area.

INTRODUCTION

Heavy machinery components offer a timely research area since the materials used for their manufacture must have specific mechanical characteristics. HSLA steels are commonly used in mining industry and they must be resistant to wear caused by the loading and unloading of minerals. The new generation of modern steels categorized as quenched and tempered low alloys abrasion resistant steels are extensively used in the areas requiring high resistance to abrasion. HSLA steels possess high strength, good impact toughness, and good weldability with certain precautions and good machinability [1]. Currently, these steels are used frequently in liners plates. Usually, welding joints of HSLA is the weakest part of the whole welding structure. Therefore, control of welding process is very important [2]. In the GMAW process influences the velocity of welding as well as cooling rate and solidification behavior of weld deposit and peak temperature of HAZ more favorably in weld fabrication to improve various joint properties [2]. Wear is a progressive destructive complex phenomena during which deterioration of surfaces occurs in industrial operation and components may fail due to, faulty design, excessive handling or mis-operation. It leads to heavy expenditure for maintenance and replacement of industrial plant equipment, causing a significant operating cost to industry [3, 4, and 5]. The pin-on-disc configuration is commonly used for wear tests in laboratories because of its simple arrangement [6]. In the pre-sent work, the effects of welding parameters, the wear behavior on the microstructure specifically on the HAZ of HSLA steel welded by multi-pass GMAW process has been investigated. The aim of this research was to evaluate wear resistance by pin on disc test and hardness on heat affected zone of HSLA steel plates.

EXPERIMENTAL

In this research, the selected material is a HSLA steel heat treated quenched and tempered, in Table I is showed the chemical composition of the selected material. The steel grade is 950A according the society of automotive engineers (SAE). These steels provide excellent results in abrasion resistance and thus prevent premature wear on the floors and walls of mining trucks.

Table I. Chemical composition of the base metal (wt. %).

C	S	Mn	P	Si	Cr	Ni	Mo	V	Nb	Ti	W
0.14	<0.002	1.15	0.012	0.34	0.10	<0.04	0.14	<0.004	<0.004	0.016	<0.0002

The weld joint was fabricated by multipass welding, through the process of welding and GMAW. The union was performed on plates 14mm thick, with a size of 101.60 mm x 101.60 mm. In order to optimize the process welding parameters were modified at each joint. Table II shows the DOE based on optimizing parameters of welding HSLA steels with thin thickness. The coupons used for the study of the microstructure, hardness and wear were cut from the joint in circular form with a diameter of 25.4 mm and a thickness of 15 mm.

Table II. Design of Experiments.

Plate designation	Voltage (V)	Current (A)	Feed rate (mm/s)
1	27	273	7.04
2	27	319	8.06
3	29	318	7.76
4	29	353	8.58
5	31	352	8.43
6	32	348	7.39
7	32	371	8.8

Microhardness measurements were performed in different areas of the joint using a microhardness Wilson Hardness Tukon 2500, with a load of 0.5 kg for the length 10s indentation. The wear tests were performed on the pin-on-disc machine according to ASTM G 99-05 standard [7]. For all samples the parameters were the same, applying a load of 58.8 N, a travel speed of 200 rpm with a radius of 9 mm at room temperature in dry sliding conditions. The test duration was 1.5 h. To complement the metallographic study of each of the areas of welding (base metal, heat affected zone and melting zone) analysis was performed for each of the solder joints proposals. Previous to observation under the optical microscope the samples were etched with Nital 5%. Furthermore, the worn surfaces of the specimens were studied using a scanning electron microscope JEOL JSM-6490LV.

RESULTS AND DISCUSSION

The succession of weld bead in a multipass welding on the base metal produces a complex HAZ, with not always adequate mechanical properties, presenting excessive softening zones, different grain sizes and formation of brittle phases. Thus HAZ presented in most cases less efficient mechanical properties of the weld. The HSLA steel heat treated quenched and tempered is constituted of martensite. The multipass welding generates a HAZ of 4 regions; the transition from one area to another is done progressively [8-13]. Figure 1 shows the HSLA steel after being GMAW welding, the specimen (a) is shown before and (b) after the wear test. The hardness profile can be observed in Figure 2, where the graphs show the hardness (a) before and (b) after the wear test in the base metal (BM), in HAZ (zone 1, zone 2, zone 3, zone 4) and in the weld metal (WM). The surfaces exposed to wear tend to develop work hardening so the hardness measurements of the different areas of the joint were taken. Comparing this way the hardness of the specimens can be observed the degree of strain hardening.

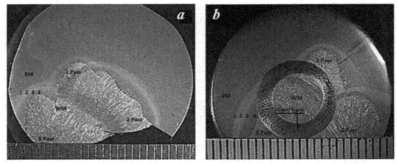

Figure 1. Microstructure of HSLA steel after GMAW process a) before, b) after the wear test.

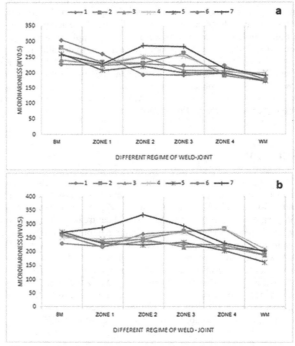

Figure 2. Hardness measurement test result in a cross section:
(a) Before and (b) After the wear test.

The wear profile of the specimen is observed in scanning electron microscopy; data from all stages of the experiments were combined to obtain the behavior of the HAZ. The wear resistance measurements were performed using the pin on disc test, determining the rate of wear and the wear rate from the mass loss experienced by the specimen. Also, measured the coefficient of friction developed during the testing. The volume loss of the worn material is determined by:

$$Volume\ loss\ (mm^3) = \frac{weight\ loss\ (g)}{density\ \left(\frac{g}{cm^3}\right)} x\ 1000m \qquad (1)$$

To calculate the rate of wear of the material is used:

$$wear\ rate\ = \left(\frac{mass\ loss\ (g)}{test\ time(min)}\right) \qquad (2)$$

The wear coefficient, calculated in terms of volume loss experienced during testing is performed according to the expression:

$$K = \frac{W}{F_N S} \qquad (3)$$

Where W is the volume lost, FN is the normal force applied to the test and S is the total linear distance [14-15]. The wear coefficient is determined from the mass loss of the coupons. The coupon 4 presents the minimum amount of mass loss, on the contrary the coupon 6 having a high mass loss. The friction coefficient has a similar tendency in all the specimens tested, but to compare the test pieces each have variation in wear behavior; this is because the welding parameters changed for each weld joint, changing the heat input and affecting hardness in each zone.

Figure 3 shows the results of the wear test, based on the coefficient calculated for each steel wear. There is a clear difference in the values obtained for the wear coefficient where it can be concluded that the specimen 4 is the one with lower wear coefficient compared to all tested specimens.

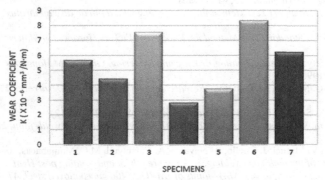

Figure 3. Wear results of the specimens tested.

CONCLUSIONS

The results obtained for the wear tests were related to hardness of the different zones and their microstructure. There is not a significant difference in the coefficient of friction measured along the wear tests for different specimens. The subzones show variation in the hardness measurements. Therefore the difference between the hardness values of a sub-zone before and after a wear test (under identical conditions) depends on its nature. The behavior of the specimens is different from the variation in the welding parameters, the specimen having the highest values along the different zones is number 6 (32 V, 348 A and 7.39 mm/s) and however presents intensive wear. While the test piece 4 (29 V, 353 A and 8.58 mm/s) has the lowest level as a wear coefficient, so it is the most resistant test. Therefore, the best parameters are used in the coupon 4.

ACKNOWLEDGMENTS

Corporación Mexicana de Investigación en Materiales (COMIMSA), Universidad Autónoma de Nuevo León, División de Estudios de Posgrado e Investigación, (FIME).

REFERENCES

1. V. Sharma, A.S. Shahi, *Materials and Design*, **53**, 727 (2014).
2. M. Mirzaei, R. Arabi Jeshvaghani, A. Yazdipour, K. Zangeneh-Madar, *Materials and Design*, **51**, 709 (2013).
3. S. Mohan, Ved Prakash, J.P. Pathak, *Wear characteristics of HSLA steel*, **252**, 16 (2002).
4. E. Sauceda Tello, *Efectos microestructurales sobre el desgaste en acero tipo AISI D2*, (1996).
5. N.A. Rodríguez Quiroga, *Efectos de los carburos en el desgaste del acero tipo AISI D2* (1998).
6. H. So, *Tribology International*, **29**, 415 (1996).
7. ASTM G 99- 05, *Standart Test Method for wear testing whit a pin on disc apparatus* (2005).
8. ASTM E 384- 11, *Standard Test Method for Knoop and Vickers Hardness of Materials* (2011).
9. American Society for Metals, *Welding Brazing and Soldering. Ohio: ASM Handbook committee*, 6 (1990).
10. O. Higuera, C. Moreno, M. Gutierrez, *Research article. Ingeniería & Desarrollo*, **27**, 151 (2010).
11. Ch. Zhang, X. Song, P. Lu, X. Hu, *Materials and Design*, **36**, 233 (2012).
12. S.N. Krishnan, V. Toppo, A. Basak, K.K. Ray, *Wear behavior of steel weld- joint*, **260**, 1285 (2005).
13. A. Martínez, V. Miguel, J. Coello, A. Navarro, A. Calatayud, M.C. Manjabacas, *Analisys of the influence of the multipass welding, welding pre a heat and welding post Heat treatments on the behaviour of GMAW joints of HARDOX400 microalloyed steel*, 47 (2011).

14. J. Górka, T. Kik, A. Czupryński, *Welding International*, **3**, 29 (2013).
15. A. García, A. Varela, J. L. Mier, C. Camba y F. Barbadillo, *Tribological study of Hadfield austenitic steels: influence of manganese on its wear response*, **46**, 47 (2010).

Mater. Res. Soc. Symp. Proc. Vol. 1766 © 2015 Materials Research Society
DOI: 10.1557/opl.2015.411

Hybrid Laser – Arc Welding Applied in Longitudinal Joints for Hydrocarbon Conduction Pipes

Raúl J. Fernández Tavitas, Rocío Saldaña Garcés, Víctor H. López Cortéz

Corporación Mexicana de Investigación en Materiales (COMIMSA)

Email: raulfdz@comimsa.com

ABSTRACT

In this paper the effect of hybrid laser arc welding on longitudinal joints for pipes of 1.27cm thick is investigated. For the investigation, an API X70 steel was welded with the HLAW process and then subjected to tensile, bending and micro hardness tests under standards for pipe manufacturing. Images of the weld seams were taken to observe the structure and size of the weld zones. Analysis was made by light microscopy to determine the phases present in the weld zones and to observe if there is a variation of grain size in the weld zones that adversely affects the mechanical properties of the API X70 steel. Results show that the mechanical properties of the joints meet the requirements for their use in pipe manufacturing; one reason is the low thickness of the weld zone that barely affects the original properties of API X70 steel. Also the presence of bainite in the microstructure of weld zones provides resistance to the joints.

INTRODUCTION

The application of hybrid laser arc welding (HLAW) has increased in recent years due to the advantages provides to joints such as good penetration, welding speed, reduced heat affected zones, reduction of process time and production costs, etc [1,2]. Nowadays welding processes for pipe joints must bring high quality and reasonable production costs to joints. Several arc processes like the submerged arc welding (SAW) or the gas metal arc welding (GMAW) are used because they are able to produce joints with high deposition rates [3], however the nature of these processes generate joints with an excessive heat input and low penetration, that lead to the use of several weld seams which cause an adversely affectation in the mechanical properties of joints. Because of this affectation, thermal treatments are used to improve the mechanical properties of the joints that causes an increment in production time and cost of pipes [4, 5]. The aim of this paper is to present the results of the hybrid laser arc welding applied in sheets of API X70 steel used in pipeline conduction. The joints are analyzed by stereoscope and light microscope to observe the weld zones and their microstructure and study their influence in mechanical properties of the joint. Also the joints are qualified under specification API 5L used in manufacturing pipes with tensile, bending and hardness tests to observe if the joints are suitable for their use in pipeline conduction.

EXPERIMENTAL

For the experiment 6 plates of API X70 steel with measures of 200 mm long, 100 mm width and thickness of 12.7 mm were used (chemical composition and mechanical properties of the API X70 steel are shown in Table 1 and 2, respectively). Two of them were mechanically prepared with a bevel of 45 degrees and a root face of 2 mm. The other four plates remained without preparation. Shielding gas was a mixture of 85% argon with 15% CO_2 [5 – 7].

Table I. Chemical composition of base metal and filler metal (wt. %).

	C	Mn	Si	S	P	Al	Nb	Cu	Cr	Ni	Ti	Ca
X70	0.04	1.5	0.21	<0.003	<0.01	0.03	0.27	0.26	0.16	0.08	0.08	0.0025

Table II. Chemical composition of base metal and filler metal (wt. %) [5 – 7]

Yield Strength (MPa)	Tensile Strength (MPa)	Elongation (%)	Absorbed Energy (-20 °C, J)
485	570	31	252

Plates without preparation were butt welded with two weld passes, one pass on the upper surface and the other on the underside (A and B). The plates with mechanical preparation were welded by butt joint with two weld passes on the upper surface (C); the welding parameters are shown in table 3.

Table III. Parameters for hybrid welding. Plates without preparation (A & B). Plates with preparation (C).

	A		B		C	
	1	2	1	2	1	2
Laser Power (kW)	3.8	3.8	3.8	3.8	3.8	2
Welding Speed (m/s)	0.005	0.005	0.005	0.005	0.23	0.14
Feed Speed (m/s)	5.88	3.54	5.88	3.54	15.36	17.1
Focal Distance (mm)	250	250	250	250	246	246
Voltage (V)	21.3	22.8	21.3	22.8	29.6	30
Amperage (A)	155	145	155	145	393	403
Filler Metal	ER70S-6	ER70S-6	E71T1	E71T1	ER70S-6	ER70S-6
Gas Flow (l/min)	8	8	8	8	8	8
Stick Out (mm)	9	9	9	9	9	9

For the experiment an HLAW equipment with power supply brand ROFIN, Nd YAG laser with 4.4 kW of maximum power with a laser beam diameter of 0.4 mm, the laser head is connected to a GMAW torch brand Fronius TPS 500 MAG. Chemical analysis of metal base was made by spectroscopy. As mentioned before the joints were qualified under specification API 5L with tensile, bending and microhardness tests. Tensile test was performed with a machine TINIUS OLSEN; the bending test was conducted with an equipment brand HOYTOM. In addition, Vickers micro hardness with an equipment INSTRON TUKON 2500. For metallographic characterization, a light microscope Olympus and a scanning electronic microscope JEOL JSM – 6490LV were used.

RESULTS AND DISCUSSION

Macroscopic cross – section of the joints are shown in Figure 1; the welding zones are indicated in the picture as well as their constitutive parts. The use of two weld passes caused the formation of subzones in the heat affected zone. The formation of the subzones is desirable because it improves the toughness of the joint [9].

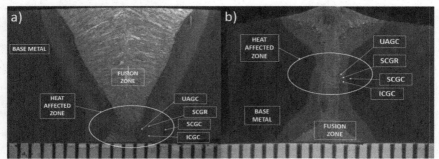

Figure 1. Macroscopic cross-section of joint welded using the HLAW process. a) Joint A and b) Joint A and B. UAGC (unaffected grain coarsened), super critically reheated grain refinement (SCGR), subcritical reheated grain coarsened (SCGC) and inter critically reheated grain coarsened (ICGC).

The microstructure of the base metal is composed of acicular ferrite and polygonal ferrite as main phase. Figure 2 shows the microstructure of the base metal where it can be seen the interlocking nature of acicular ferrite that coexists with other ferrite morphologies such as polygonal and quasi polygonal ferrite, Widmastattën ferrite and allotriomorphic ferrite; acicular ferrite also may grows from inclusions or from another preexisting ferrite plates as seen in figure 3. The presence of this microstructure is typical in steels used in oil pipelines because it improves their toughness and impact resistance [4, 9].

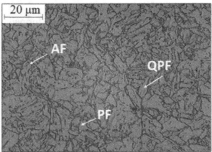

Figure 2. Typical microstructure of an API X70 steel, the microstructure is composed by acicular ferrite – AF, polygonal ferrite – PF, quasi polygonal ferrite – QPF, Widmastattën ferrite – FW.

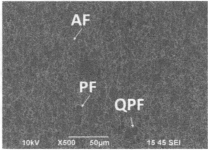

Figure 3. Base metal microstructure, image taken by SEM, AF- acicular ferrite, PF- polygonal ferrite and QPF- quasi polygonal ferrite.

When the heat affected zone was analyzed in light microscope, the microstructure presented in this zone consists in high volume fractions of sheaf like bainitic structure with martensite and austenite aggregates growing at grain boundary; it also contains some acicular ferrite (figure 4). This structure may be produced by the high cooling rate provided by the laser welding that transforms some austenite grain into martensite, however the interaction of the GMAW reduces the cooling rate and this tends to transforms austenite into bainite and acicular ferrite [10,11]. Figure 5 shows the bainitic structure using the scanning electron microscope. This microstructure improves the tensile strength and impact resistance to the joints.

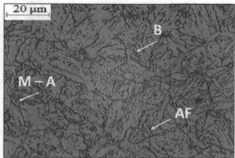

Figure 4. Heat affected zone of joint without mechanical preparation (A), welded with the hybrid process, image taken by light microscope.

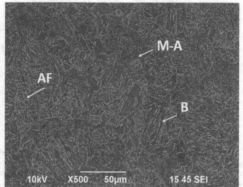

Figure 5. Microstructure of heat affected zone, image taken by SEM. AF- acicular ferrite, B-Bainite and M-A – martensite and austenite aggregates.

Also the microstructure of the fusion zone consists of bands of ferrite with martensite in major proportion with a little percentage of bainite and acicular ferrite [9], as it can be seen in Figures 6 and 7. This microstructure is obtained because of the high cooling presented in this zone after the welding process [1].

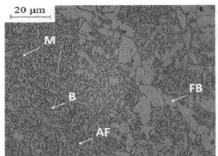

Figure 6. Fusion zone of joint with mechanical preparation welded with HLAW process. M – Martensite, B – Bainite, AF – Acicular Ferrite and FB – Ferrite Bands.

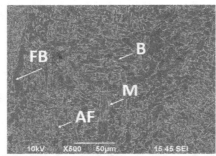

Figure 7. Microstructure of fusion zone. Image taken by SEM.

As mentioned above the tensile tests were conducted under specification API 5L [8]. The results of tensile tests performed for HLAW joints show an improved resistance greater than the original properties; this is produced because the weld seam is little so it does not affects the properties of the steel, on the contrary it improves the tensile strength of the steel. Also during welding process, the heat input of the laser focuses more heat into weld pool and the heat affected zone gets reduced, the affected area is minimal and it causes a minimum impact on steel properties. Table IV shows the results of tensile tests [1].

Table IV. Microstructure of fusion zone. Image taken by SEM.

N°	Joint	Strength	Fracture zone
1	A	630 MPa	Base Metal
2	B	627 MPa	Base Metal
3	C	624 MPa	Base Metal

The specification establishes that the steel must have at least 300 MPa of tensile strength and also the fracture must occur in the base metal. The requirements established by the specifications were accomplished. During tensile test a ductile fracture occurs in the samples; this phenomena can be seen in Figure 8; its structure is typical for a ductile fracture by dimpling and distortion of crystal planes [12].

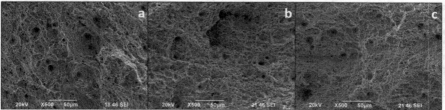

Figure 8. Image by SEM that shows the typical structure of ductile fracture. a) Fracture of joint A, b) Fracture of joint B and c) Fracture of joint C.

Bending tests of the welded plates were performed with a bending angle of 180°. The bended plates did not show defects in the joints like pores or discontinuities [8]. Micro hardness tests were performed under API 5L specification [8]. The results of the micro hardness test are shown in Figure 9. The hardness of base metal is the same for the three joints but there is an increment of hardness in the heat affected zone due the presence of bainite and martensite. However the hardness of joint C is smaller than in joints A and B. This is caused for the second weld seam used for joint C; the weld seam act as a thermal cycle that reduces the hardness of the fusion zone [5].

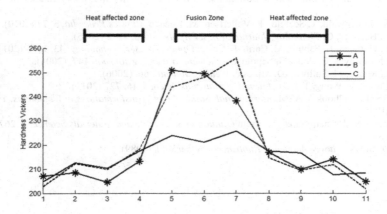

Figure 9. Results of hardness test under specification API 5L.

CONCLUSIONS

In this study, microstructure, hardness, toughness and fracture morphology of API X70 steel with HLAW process were studied. The analysis indicates that the HLAW is able to produce joints that meet the requirements for their use in oil pipelines and also aggregates advantages like savings in time and cost of production; it also achieves a correct penetration with a slim weld zone. The microstructure of the heat affected zone consists of bainite with martensite – austenite aggregates. This microstructure is caused by the high welding speed and the accelerated cooling rate. The cooling rate at the time of the joints changes the original structure of the steel, however due to the laser welding process allows a low affectation of the base metal and can thus preserve the original properties even in weld zone. The microstructure obtained improves the mechanical properties of the joints. It is highly recommended to continue with the investigation to realize more proves that confirms the good quality and effectiveness that the HLAW process can bring to pipe welding.

ACKNOWLEDGMENTS

Research sponsored by Corporación Mexicana de Investigación en Materiales S.A., Consejo Nacional de Ciencia y Tecnología, Centro Tecnológico AIMEN.

REFERENCES

1. A. Shahnwaz, M. Khan, *International Journal of Engineering and Innovative Technology*, **1**, 97 (2012).
2. M. Fersini, G. Demofonti, S. Sorrentino, E. Mecozzi, *Welding International*, **23**, 450 (2009).
3. S. Grünenwald, T. Seefeld, F. Vollertsen, M. Kocakb, *Physics Procedia*, **5**, 77 (2010).
4. C. Bagguer, O. Flemming, *Journal of Laser Applications*, **17**, 2 (2005).
5. E. Le Guen, E. Fabbro, M. Carin, F. Coste, *Optics & Laser Technology*, **43**, 1155 (2011).
6. M. El Rayes, C. Walz, G. Sepold, *Supplement to Welding Journal*, **147**, (2004).
7. E. Retzel, M. Sullivan, D. Mikesic, *Welding Journal*, **66**, (2006).
8. Ch. Li, Y. Wang, T. Han, *Journal of Materials Science*, **46**, 727 (2011).
9. ASM Handbook 6, ASM *International. Solid – State Transformations in Weldments*, 191 (1995)
10. Babu, Sudaranam Suresh, Current *Opinion in Solid State and Materials Science*, **8**, 267 (2004).
11. D. Broek, *Kluwer Academic Publishers, Netherland*, (1989).

Mater. Res. Soc. Symp. Proc. Vol. 1766 © 2015 Materials Research Society
DOI: 10.1557/opl.2015.412

Study of GMAW Process Parameters on the Mechanisms of Wear in Contact Tips C12200 Alloy

Luis A. López, Gladys Y. Perez, Felipe J. Garcia, Víctor H. López

Corporación Mexicana de Investigación en Materiales (COMIMSA), Calle ciencia y tecnología
No. 790, Col. Saltillo 400, cp. 25290, Coahuila, México.
E-mail: alberto.lopez@comimsa.com

ABSTRACT

This paper focuses on the impact of process parameters of gas metal arc welding (GMAW) on the mechanisms of fail and wear present in the contact tips (CT), component located in the welding gun, when high strength low alloy (HSLA) steel is welded with ER70S - 0.045" copper coated electrode in manual mode. By means of chemical analysis the alloy was identified as C12200. It was also identified that the maximum temperature reached by the CT is 850° C. 30 samples were obtained that had different lifetime, which were analyzed by stereoscope and its behavior against wear was determined by using an equation of relative wear. Microstructural changes as recrystallization and grain growth undergone by these CT were also evidenced by light microscopy. In addition the changes in their mechanical properties such as decrease in their hardness to about of half that initially had. Finally some significant samples were analyzed by scanning electron microscopy (SEM); microanalysis was used to identify the exchange of matter leaving from the electrode in the CT and spatter into the hole of the component.

INTRODUCTION

GMAW is a welding process that uses an arc between a continuous filler metal electrode and weld pool, this process is performed under the protection of a gas [1]. The process is commonly used in large factories as it is ideal for welding applications in high volume production for having high deposition rates and high welding speeds [2]. The CT is a key component in the GMAW gun because it has the function of positioning the electrode and supply it electric current efficiently to produce quality welds. This component is usually made of copper alloys because this material has good electrical and thermal conductivity and a high melting point. Characteristics necessary to be exposed to wear by the sliding contact between the contact tip and the welding wire at high temperature and electrical erosion. The actual contact area between the wire and the contact tip bore is significantly smaller than that result from the geometric apparent contact area [3] When the contact tip deteriorates or fails different problems are generated: false contacts, electrical disruption, arc instability, obstruction in the passage of the electrode, among others [4]; leading to defects and poor weld appearance this in turn implies rework, successive work stoppages by component replacement and downtime thus affecting the productivity of companies using this process.

EXPERIMENTAL

In this work, a chemical analysis was performed on the CT and the elements that compose it are shown in Table 1. Once identified the alloy the mechanical properties were obtained which are shown in Table 2 [5]. In Figure 1 shown a new contact tip and its dimensions are listed in Table 3. After that working conditions in the factory were replicated, the operating parameters shown in Table 4 a thermocouple was installed in the component as shown in Figure 2 in order to identify the temperature reached by the CT during its service life. Where the welding machine used was a DeltaWeld 652.

Table I. Chemical composition of the copper alloy C12200 (wt %).

Wt %	Cu	Al	P	Si	Cr	Mg	Mo	Mn	Zn	Ti	V
C12200	Balance	0.001	0.020	<0.001	<0.001	<0.001	<0.001	<0.001	0.061	<0.001	<0.001
	Zr	Fe	Sn	Pb	Sb	Co	Cd	Ni			
	<0.001	0.012	0.002	0.004	0.001	<0.001	<0.001	0.001			

Table II. Mechanical properties of the copper alloy C12200.

Base Metal	Yield Strength	Ultimate Tensile Strength	Elongation (%)	Hardness (HV)
C12200	310 MPa	345 MPa	10%	108 HV

Figure 1. Appearance of a new contact tip.

Table III. Dimension of the contact tip.

Length	Outer diameter	Inner diameter	Diameter of wire used	Difference between electrode and hole
34 mm	7.6 mm	1.4 mm	1.14 mm	0.26 mm

Table IV. Welding Parameters.

Voltage	Current	Shielding gas	Flow of gas	Wire feed rate	Welding speed	Extension length
32 V	330 A	Ar + 10%CO_2	45 CFH	16.5 m/min	510 mm/min	25 mm

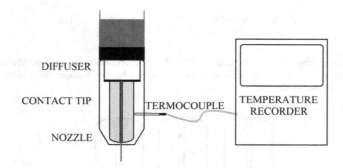

Figure 2. Schematic drawing of temperature measurement of contact tip during GMA welding

30 tips were identified and careful monitored in workstations that have a high consumption of CT in order to determine the arc time that each of them works. Then the surface of the samples was cleaned with a fine grinding with 600 grit SiC paper and water like lubricant then ultrasonic cleaning was performed for a period of 50 min. Under these conditions the wear measurements were performed as shown in Figure 3 with the support of analyzer images software in images taken by Olympus stereoscope SZX10.

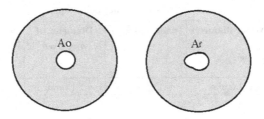

Figure 3. Schematic drawing of the areas A_0 y A_t.

The method consists in measure the area of the hole before and after welding as shown in figure 3 and the wear in weight loss % was determined: subtracting the initial area A_0 to worn area A_t divided by the initial area A_0 and multiplying by 100 as shown equation 1 [7] used to determine the wear in weight loss %.

$$Wt(\%) = \frac{A_t \cdot A_0}{A_0} \times 100 \qquad (1)$$

After calculating the wear on the contact tips by the aforementioned method, the samples were cut longitudinally as shown in Figure 4, to take images by stereoscope and to identify the change in the surface of the hole.

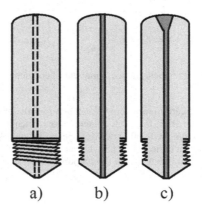

a) b) c)

Figure 4. Schematic drawing of the new contact tip a) complete b) cut longitudinally c) after "t" min of arc time cut longitudinally.

Then these halves were sectioned transversely, mounted in Bakelite grinded and polished to get observation by microscope, the fine grinding were performed with 400 and 600 grit SiC paper, and the polishing in cloth with diamond paste to polishing of 1 microns and colloid silica of 0.04 microns, finally were etched with potassium dichromate $K_2Cr_2O_7$ by 6 seconds to identify grain boundaries [8] as shown the Figure 5. And thereby determine the microstructural changes undergone by the material.

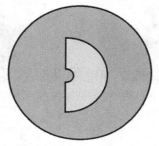

Figure 5. Schematic drawing of the metallographic preparation of the CT.

To support the results obtained in the metallographic hardness tests were performed in Figure 6 shown as were carried out 10 measurements of hardness in each CT.

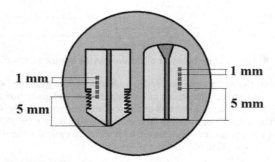

Figure 6. Schematic drawing of the measurements of hardness.

Finally analyze CT by SEM looking to identify the wear mechanisms in the component during its lifetime.

RESULTS AND DISCUSSION

In Figure 7 two samples are compared both shown in top view. Figure 3a shown the CT before welding and Figure 3b shown CT after welding "265" minutes, where the changes that occurs in the surface of the component after work are clearly observed. Some studies [9, 10] have found that a relatively small region at the bore outlet takes most of the damage by wear. The results obtained from the temperature tests are shown in Figure 8. Where the range displayed is only the first 180 seconds of the test, after this time the temperature of the component stabilized [11] at around 850 °C.

Figure 7. Area of exit hole a) Before welding b) After "265" minutes welding

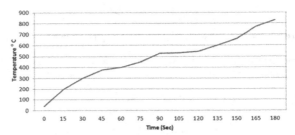

Figure 8. Temperature behavior of the contact tip as time function

The time worked by the contact tips exposed to high temperature generated a change in its microstructure, Figure 9 shown the phenomena of recrystallization and grain growth the grain growth in their microstructure after 312 and 608 minutes of arc time respectively; since the mechanical properties of the material are linked to their microstructure these were also affected, the Figure 10 shows a graph of the hardness against arc time which has initial value of 130 Hv and drops to values of 65 Hv from 200 to 600 minutes of arc time.

Figure 9. Micrographs comparing the microstructure of the material a) with no arc time, b) after 312 min of arc, c) after 608 min of arc

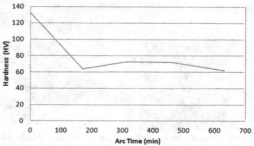

Figure 10. Changes in the hardness of the contact tip.

The results of wear measured in Figure 11 was obtained from equation 1 and shows a normal trend of wear to increase, this trend is not completely uniform, because the experiment was done in a manual process where a welder was working with the component. In the graph the relationship between the wear and the arc time is clearly identified. Although it is noteworthy that the wear is not as much as that generated when using copper uncoated electrode, the rate of wear of this type of electrode is 10 times the copper coated electrode [12].

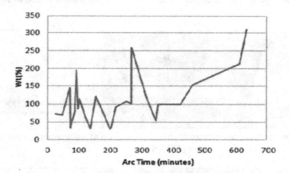

Figure 11. Wear in weight loss (%) of the contact tip in function of arc time

Figure 12 a shown the bore surface of a new CT cut longitudinally in which the surface of the hole is obtained with the manufacturing process that looks a smooth surface with good finishing which means good surface for electrical contact with the electrode. In Figure 12b is perceived the wear generated by the electrode in the tip of the CT also shows the changes in the surface of the bore that looks similar to valleys and crests not favorable conditions to the correct flow of current from de CT to the electrode since it has spatter of steel, copper carbonized material and a rough surface which in addition to compromising the flow of the current, can be cause that the electrode is get stuck in the CT. Some researches [13, 14] have suggested that electrical erosion is the key wear mechanism.

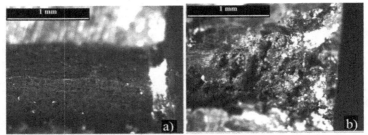

Figure 12. Sectional view of the surfaces of the CT bore a) before welding b) after welding.

Finally some significant samples were analyzed by SEM in which were identified small mounds of steel into the bore of the CT and even found exchanges of matter from the electrode to the intermediate zone of the CT, in these areas were conducted EDS microanalysis confirming that the zones shows a high content of Fe. The exchange of material in Figure 13 is given by the electrode exchange to the contact tip, as well as embedded splatter inside the hole. While in Figure 14 the exchange of electrode to the contact tip is located at 10.5 mm from the end that is closest to the weld pool, in both samples was made microanalysis in order to confirm the presence of steel. It is noteworthy that these results are consistent with those found in other studies [4].

Figure 13. Steel transfer from the electrode to the contact tip a) Stereoscopic image b) SEM image c) EDXS spectrum

Figure 14. Evidence of exchange electrode material to the contact tip
a) Stereoscopic b) SEM image c) EDXS spectrum.

CONCLUSIONS

The maximum temperature of the component was identified as 850 °C under the specified parameters listed in table 4. Microstructural changes undergone by the component were identified. As the recrystallization and the growth grains while the hardness decrease from 130 Hv to 65 Hv and thereby the wear resistance also decreased, because the hardness of the material is one of the main characteristics to consider in the study of the resistance against wear of a material. Also the measuring indirectly of wear as weight loss suffered by the component allowed to know the behavior of the CT to the phenomenon of wear, which is not one of the main reasons why the CT is replaced. The exchange of material from the electrode to the inside of the bore in the CT was identified with the assistance of SEM, these exchanges happens by the effect of arcs generated between the surfaces of these two components; this means that the air present between these two surfaces inside the hole is ionized and an electric arc is generated, causing melting of the metals involved. In other sample analyzed was identified spatter in the inner of bore on the end that is closer to the arc, and also were identified the changes in the condition of the surface of the bore of the CT as mentioned earlier a smooth surface and uniform become a rough surface with the impression that has crests and valleys. These latter effects are detrimental for the correct flow of electricity and the free passage of the electrode through the CT. Because cause that the component does not meet with its two main functions that are supply electric current to the electrode and feeding the wire efficiently; being this the main reason why the component is changed. The durability of these CT is related to the effect of the parameters on the working conditions and the physical and metallurgical properties of the material from which the component is manufactured. The effects mentioned above are a natural consequence of the work parameters analyzed, however it is concluded that the wear on the component is not the main reason why the component is replaced. Also since this is a manual process where the weld is applied by a welder is recommended to have an efficient care when removing the remains of splatter of the end of CT, to prevent that the orifice becoming clogged.

REFERENCES

1. R.L. O'Brien, *Manual de soldadura, AWS*, 8th Edition, 110 (1996).
2. A.D. Althouse, C.H. Turnquist, *Modern Welding*, 11th Edition, 233 (2013).
3. D. Kristóf, L. Németh, *Design Fabrication and Economy of Metal Structures, Budapest, Hungary*, 489 (2013).
4. J. Villafuerte, *Welding journal*, **78**, 29 (1999).
5. Cooper Development Association Inc. *Properties of Wrought and Cast Copper Alloys. A Copper Alliance* (http://www.copper.org/resources/properties/db/basic-search.php) Accessed 30 Jul (2013).
6. N.H. Kim, K.H. Kim, H.J. Kim, H.S. Ryoo, J.H. Koh, *Journal of KWS*, **22**, 43 (2004).
7. N.H. Kim, K.H. Kim, H.S. Ryoo, *Development of Cr- Cu Contact Tips for GMAW Welding, Journal of KWS*, **23**.
8. G.F. Vander Voort, *Metallography, principles and practice*, McGraw Hill New York, ASM International, USA (1984).
9. J. Villafuerte, *Tregaskiss R y D Doc. 2001* (1997).
10. T. Yamada, O. Tanaka, *Welding Journal*, **66**, 35 (1987).
11. B.G. Adam, T.A. Siewert, T.P. Quinn, D.P. Vigliotti, *Welding Journal*, 37 (2001).

12. H. Shimizu, Y. Yokota, M. Mizuno, T. Kurokawa, *Science and Technology of Welding and Joining,* **11**, 94 (2006).
13. H. Their, H. Polrolniczak, S. Schreiber, *Schweissen und Schneiden,* **47**, no.5, (English translation of text and captions pp. 356, 358, 360, 362, 365.E88-E90).
14. V.G. Degtyarev, M.P. Novikov, N.M. Voropai; Paton, *Welding Journal 3,* **4**, 290 (1991).

Characterization of Steels Used in the Oil Industry

Mater. Res. Soc. Symp. Proc. Vol. 1766 © 2015 Materials Research Society
DOI: 10.1557/opl.2015.413

Biocorrosion of Bacterial Inoculation on the API X52 Pipeline Steel

M. Amaya[1], V. L. Reyes-Martínez[1], J. M. Romero[1], L. Martinez[2] and R. Perez[3]

[1] Instituto Mexicano del Petróleo, Eje Central Lázaro Cárdenas Norte 152, Col. San Bartolo Atepehuacan, México D. F., C.P. 07730, México.
E-mail: jromero@imp.mx

[2] Instituto de Ciencias Físicas, Universidad Nacional Autónoma de México, Ap. Postal 43-8, Cuernavaca, Morelos 62251, México.
E-mail: lmg@corrosionyproteccion.com

[3] Centro de Física Aplicada y Tecnología Avanzada, Universidad Nacional Autónoma de México, Boulevard Juriquilla 3001, Santiago de Querétaro, Qro., 76230, México.
E-mail: ramiro@fata.unam.mx

ABSTRACT

The role of the initial bacterial inoculates on the biocorrosion of API X52 pipeline steel coupons was evaluated by electrochemical noise technique. The experiments were performed under laboratory conditions using an aerobic bacteria identified as *Achromobacter xylosoxidans*. Inoculations in the interval strain of $1x10^4 - 1x10^8$ CFU/ml were evaluated. Environmental scanning electron microscopy (ESEM) analysis was carried out to evaluate the corrosive effects induced on the API X52 electrodes. The results show that all corroded surfaces show sites of localized corrosion, however, the density of de sites of localized corrosion have different grades depending of the initial inoculation used during the experiments. The maximum density sites of localized corrosion were obtained in the experiments with $1x10^5$ CFU/ml. From inoculates of $1x10^6$ CFU/ml the density sites of localized corrosion diminished constantly. The results show that with inoculates over $1x10^6$ CFU/ml, the oxygen demand for the bacterial strain limits the presence of oxygen available into the metallic surface to maintain the corrosion reactions. The results were supported by the EDX analysis of the corrosion products formed on the metallic surfaces where the oxygen peaks diminished as the bacterial inoculation increases.

INTRODUCTION

Commonly in the microbiological induced corrosion studies under laboratory conditions it selected inoculates from $1x10^3 - 1x10^9$ CFU/ml (Colony Forming Units/milliliter) [1-9]. Independently of the bacteria strain analyzed and the flow conditions selected, it appears that the amount of culture medium and the initial inoculums was arbitrarily selected. For this reason in the present study it was evaluated the effect of the initial inoculum of an aerobic bacterial strain under laboratory conditions. Several electrochemical experimental measurements were performed in an electrochemical cell containing culture media inoculate whit a bacterial strain using as working electrodes coupons of the API 5L X52 pipeline steel.

EXPERIMENTAL

Several cylindrical coupons were machined from a commercial pipeline steel (API 5L X52). Three identical electrode cells were constructed using the coupons with 10 mm in length and 8 mm in diameter. The electrode cells consist of three cylindrical coupons embedded in polyester resin. The three electrode cells were mechanically polished with increasingly fine grid SiC paper, grade 400 and 600 and then with alumina (Al_2O_3) of 3μm diameter. All three-electrode cells were rinsed in distilled water and acetone (CH_3COCH_3), dried in warm air, and maintained in a recipient containing silica to prevent oxidation after the electrochemical experiments.

A bacterial strain isolated from the biofilm formed on a coupon exposed inside of a pipeline that transports seawater used in an offshore injection system located in the Gulf of Mexico was used during the experiments The bacterial strain selected was identified as *Achromobacter xylosoxidans*. The isolation procedure and identification are described elsewhere [10].

Previously to the electrochemical experiments, laboratory conditions were set to initiate and sustain growth conditions of bacterial strain in a batch culture media. The recommended formulation is described elsewhere [11].

The electrochemical cells used during the experiments, consisting of glass flasks containing 200 ml of culture medium (pH adjusted at 7.2) and a three electrode cell immersed in this culture media inoculated with the an aerobic bacterial strain. Before the electrochemical experiments, all containers were sterilized in an autoclave and the three electrode cells were disinfected with benzalkonium chloride and UV light. A culture media without inoculate was used as control, and five culture media with inoculates $1x10^4$, $1x10^5$, $1x10^6$, $1x10^7$, and $1x10^8$ CFU/ml were evaluated. Current noise and potential noise were recorded using an auto zero-resistance ammeter (ZRA) measurement system coupled with a desktop computer. Time noise series consisted of blocks of 1,024 readings were taken during 1 s, and collected during 95 h.

After the experiments, the corroded products formed on the surface electrode cells were analyzed by environmental scanning electron microscopy (ESEM) equipped with an energy-dispersive X-Ray (EDX). Then the corroded products were cleaned in a solution containing HCL (774 ml), Sb_2O_3 (20 g), and $SnCl_2$ (50 g). The cleaned surfaces were analyzed to determine the corrosion damage induced by the bacterial strain.

RESULTS AND DISCUSSION

Typical corroded surfaces after the electrochemical experiments are shown in Figure 1. Practically all corroded surfaces show sites of localized corrosion, however, the density of sites of localized corrosion have different grades depending of the initial inoculation bacterial strain used during the experiments. The maximum density sites of localized corrosion were obtained in the experiments with $1x10^5$ CFU/ml (Figure 1c). From inoculates of $1x10^6$ CFU/ml the density sites of localized corrosion diminished constantly, (Figure 1d, 1e, and 1f).
.

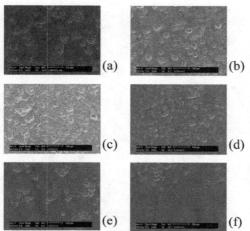

Figure 1. Typical corroded surfaces generated after the experiments (a) control, (b) $1x10^4$, (c) $1x10^5$, (d) $1x10^6$, (e) $1x10^7$, and (f) $1x10^8$ CFU/ml, respectively.

Mechanisms that inhibit the corrosion in the presence of aerobic bacterial strains are described elsewhere [4-6, 8-9, 12-14]. In this case, the results show that with inoculates over $1x10^6$ CFU/ml, the oxygen demand for the bacterial strain limits the presence of oxygen available into the metallic surface to maintain the corrosion reactions.

The EDX analysis of the corroded products (Fig. 2) show that the oxygen peaks diminished as the bacterial inoculation increases. On the other hand, the morphology of localized corrosion sites induced by aerobic bacteria in the present work were similar to the induced morphology by sulphate reducing bacteria on carbon steel reported by Romero [9].

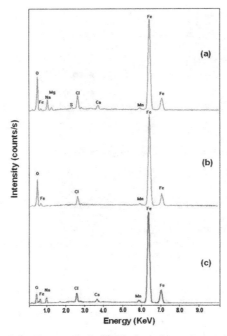

Figure 2. Energy dispersive X-ray analysis (EDX) plots of corrosion products (a) control, (b) $1x10^5$ CFU/ml, and (c) $1x10^7$ CFU/ml, respectively.

After the electrochemical experiments the potential and current noise data were used in order to obtain the pitting index (PI). The procedure to calculate the pitting index is described elsewhere [15]. Eden [16] has reported the type of corrosion expected depending on the interval of values adopted for PI: 0.001<PI<0.01 (uniform corrosion), 0.01<PI<0.1 (mixed corrosion), and 0.1<PI<1.0 (localized corrosion). A plot with the frequency of PI values generated during the experiments in the interval 0.1<PI<1.0 for all experimental conditions is shown in figure 3. The PI frequency values in the interval 01>PI>1.0 shows a Gaussian shape curve as the initial inoculate are increased. Compared with the control experiment, inoculation of $1x10^4$ and $1x10^5$ CFU/ml increase the frequency of PI, but with inoculations of $1x10^6$, $1x10^7$ and $1x10^8$ CFU/ml, PI tends to diminish. Notice that with inoculations of $1x10^7$ and $1x10^8$ CFU/ml, the frequencies of PI are lower as compared with the PI frequency measured in the control experiment. Several researchers [1, 4-6, 8] found inhibited corrosion in carbon steel in the presence of aerobic, facultative anaerobic, hydrogen-oxidizing or microaerobic bacterial strains. All these inhibit corrosion results were obtained with inoculations over $1x10^7$ CFU/ml. This inhibited corrosion was observed in short-time or long-time experimental conditions. Our results with inoculates of $1x10^7$ and $1x10^8$ CFU/ml are consistent whit these last results.

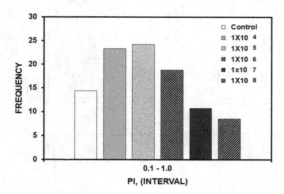

Figure 3. Frequency of PI values in the interval 0.1>PI>1.0 obtained after electrochemical noise experiments.

On the other hand, the PI values obtained at the final of each experiment should reflect the electrochemical conditions in this period time. Consequently, the corroded surfaces should reflect the final corrosion mechanism that control this final period time. The final PI values measured in each experiment were 0.75, 0.90, 0.90, 0.12, 0.07 and 0.02 for the control, $1x10^4$, $1x10^5$, $1x10^6$, $1x10^7$ and $1x10^8$ CFU/ml, respectively (Fig. 4). These results suggest that the main corrosion mechanism in the first four experiments was the localized corrosion, while in the last two experiments the main corrosion mechanism was controlled by a mixed corrosion mechanism.

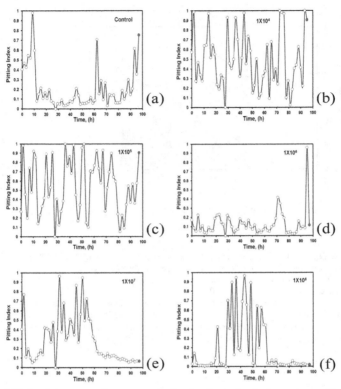

Figure 4. Pitting index versus time generated after the electrochemical experiments (a) control, (b) $1x10^4$, (c) $1x10^5$, (d) $1x10^6$, (e) $1x10^7$, and (f) $1x10^8$ CFU/ml, respectively.

CONCLUSIONS

All corroded surfaces show sites of localized corrosion, however, the density of de sites of localized corrosion have different grades depending of the initial inoculation bacterial strain used during the experiments. The maximum density sites of localized corrosion were obtained in the experiments with $1x10^5$ CFU/ml. From inoculates of $1x10^6$ CFU/ml the density sites of localized corrosion diminished constantly. The results show that with inoculates over $1x10^6$ CFU/ml, the oxygen demand for the bacterial strain limits the presence of oxygen available into the metallic surface to maintain the corrosion reactions. The results were supported by the EDX analysis of the corrosion products formed on the metallic surfaces where the oxygen peaks diminished as the bacterial inoculation increases. On the other hand, the PI measurements obtained after electrochemical experiments show good correlation with de corrosion mechanisms observed on the metallic surfaces.

ACKNOWLEDGMENTS

The authors acknowledge the SEM technical support of C. Angeles-Chavez and P. Morales-Gil.

REFERENCES

1. S.K. Jigletsova, V.B. Rodin, V.S. Kobelev, N.V. Aleksandrova, G.E. Rasulova, V.P. Kholodenko, *Applied Biochemistry and Microbiology*, **36**, 550 (2000).
2. H. Liu, L. Xu, J. Zeng, *British Corrosion Journal*, **35**, 131 (2000).
3. E.E. Roden, M.M Urrutia, C.J. Mann, *Applied and Environmental Microbiology*, **66**, 1062 (2000).
4. V.B. Rodin, S.K. Jigletsova, V.S. Kobelev, N.A. Akimova, N.V. Aleksandrova, G.E. Rasulova, V.P. Kholodenko, *Applied Biochemistry and Microbiology*, **36**, 589 (2000).
5. Y. Tanji, T. Itoh, T. Nakano, K. Hori, H. Unno, *Corrosion*, **58**, 232 (2000).
6. M. Dubiel, C.H. Hsu, C.C. Chien, F. Mansfeld, D.K. Newman, *Applied and Environmental Microbiology*, **68**, 1440 (2002).
7. R.P. George, D. Marshall, R.C. Newman, *Corros. Sci.,* **45**, 1999 (2003).
8. G. Gunasekaran, S. Chongdar, S.N. Gaonkar, P. Kumar, *Corros. Sci.,* **46**, 1953 (2004).
9. M. De Romero, Z. Duque, L. Rodríguez, O. De Rincón, O. Pérez, I. Araujo, *Corrosion*, **61**, 68 (2005).
10. J. Jan Roblero, J.M. Romero, M. Amaya, S.L. Borgne, *Appl. Microbiol. Biotechnol.,* **64**, 862 (2004).
11. NACE Report no. 54281, "Review of Current Practices for Monitoring Bacterial Growth in Oilfield Systems", (Houston, TX: NACE, 1990).
12. A. Jayaraman, J.C. Earthman, T.K. Wood, *Appl. Microbiol. Biotechnol.,* **47**, 62 (1997).
13. K.M. Ismail, T. Gehrig, A. Jayaraman, T.K. Wood, K. Trandem, P.J. Arps, J.C. Earthman, *Corrosion*, **58**, 417 (2002).
14. B. Little, R. Ray, *Corrosion*, **58**, 424 (2002).
15. J.M. Romero, J.L. Villalobos, E. Sosa, M. Amaya, *Corrosion*, **60**, 689 (2004).
16. D.A. Eden, D.G. John, J.L. Dawson, *"Corrosion Monitoring"* International Patent WO 87/07022 (World Intellectual Property Organization, Nov. 19, 1997).

Mater. Res. Soc. Symp. Proc. Vol. 1766 © 2015 Materials Research Society
DOI: 10.1557/opl.2015.414

Electrochemical Kinetic of a Low Carbon Steel in Seawater at Different Flow Speed

A. Carmona[1], R. Orozco-Cruz[1], E. Mejía-Sánchez[1], A. Contreras[2], R. Galván-Martínez[1*]

[1] Unidad Anticorrosión, Instituto de Ingeniería, Universidad Veracruzana, Av. S.S. Juan Pablo II s/n, Fracc. Costa Verde, C.P. 94294, Veracruz, México.
*E-mail: rigalvan@uv.mx

[2] Instituto Mexicano del Petróleo, Eje Central Lázaro Cárdenas Norte No.152, Col. San Bartolo Atepehuacan, Del. Gustavo A. Madero, C.P. 07730, México.

ABSTRACT

An electrochemical impedance spectroscopy (EIS) corrosion study of API X70 steel was carried out in synthetic seawater with different rotation speeds using a rotating cylinder electrode (RCE) to control the hydrodynamic conditions at room temperature, atmospheric pressure and 24 h of exposure time. A superficial analysis through a scanning electron microscope (SEM) was used to analyze the corrosion type. The rotation speed used was 0 rpm (static condition), 1000, 3000 and 5000 rpm (turbulent flow). The results show that the turbulent flow conditions affect directly the corrosion rate (CR) of the steel, because all values of the CR under turbulent flow conditions are higher than the CR values at static conditions. In addition, it is important to point out that at turbulent flow conditions, the CR increased as the rotation speed also increased. The morphology of the corrosion in all experiments was localized corrosion.

INTRODUCTION

The majority of processes in the oil industry involve the hydrocarbons transport, where this hydrocarbon could be a mix of the liquid or gas with water or humidity and it is transported by steel pipeline [1]. The most common steels used in the hydrocarbons transport belong to the 5L specification of the American Petroleum Institute (API) specifically the steels X52, X60 and X70 [2]. All studies are focused mainly in the API steel corrosion under static conditions, but it is important to point out that the principal regimen in the industry and specifically in pipeline of hydrocarbons transport is the turbulent flow; however, few corrosion studies in controlled turbulent flow conditions are available. With the increasing necessity to describe the corrosion of metals in turbulent flow conditions, some laboratory hydrodynamic systems have been used with different degrees of success [3]. The rotating cylinder electrodes (RCEs), pipe segments, concentric pipe segments, submerged impinging jets and close-circuit loops have been used as hydrodynamic systems in order to study the corrosion phenomenon under turbulent flow conditions [3-6]. The use of RCE, as a laboratory hydrodynamic test system, has been gaining popularity in corrosion studies under turbulent flow conditions [6-8]. This popularity is due to its characteristics such as, its operation mainly in turbulent flow conditions, its well-defined hydrodynamics, ease assembly and disassembly and smaller volume of fluid used [9-11]. In order to analysis the corrosion process of the steel sample, the electrochemical impedance spectroscopy (EIS) was used. This technique is a powerful tool, where the conventional current direct techniques have limitations. EIS measurements are carrying out in alternate current and it provides results about the metal-electrolyte interface and its electrical components [12]. This paper presents the results of the electrochemical corrosion techniques using a RCE in order to control the hydrodynamic conditions to the corrosion process of the steel pipeline sample immersed in synthetic seawater with different speed rotation.

EXPERIMENTAL

Materials
Cylindrical working electrodes made of API X70 steel were used in all experiments. The total exposure area of the working electrodes was 3.45 cm^2. Prior to each experiment the steel working electrodes were polished up to 600 grit SiC paper, cleaned with destilated water and degreased with acetone.

Test solution
The solution used in all electrochemical tests was synthetic seawater and it was prepared according to the standard of ASTM D1141. The exposure time in all corrosion tests was 24 h at room temperature and atmospheric pressure (Veracruz Port, Mexico). In order to control the hydrodynamic conditions, a RCE was used and the rotation speed of this RCE was 1000, 3000 and 5000 rpm. It is important to point out that the electrochemical measures were also carried out at static conditions.

Electrochemical techniques
All the electrochemical experiments were performed with a potentiostat/galvanostat. These electrochemical measurements were: Corrosion potential (E_{corr}) and electrochemical impedance spectroscopy (EIS).
In all EIS tests, the frequency range used was 0.01 Hz to 10 kHz with a 0.001 V of amplitude. 10 points per decade of frequency were recorded.
In order to get the morphology of the corrosive process, some exposed steel samples were selected in order to made a superficial analysis using a scanning electron microscope.

Experimental set-up
An air-tight three-electrode electrochemical glass cell was used; working, reference and auxiliary electrode. The reference electrode (RE) used was a saturated calomel electrode (SCE) and a sintered graphite rod was used as auxiliary electrode (AE). In order to minimize the effect of the solution resistance a Lugging capillary was used.

RESULTS AND DISCUSSION

Steel used
The steel sample used as working electrode was made of API 5L X70 pipeline steel with 36" in diameter and 0.902" in thickness. Table I shows the chemical composition of this steel.

Table I. Chemical composition of API X70 pipeline steel (wt. %).

C	Mn	Si	P	S	Cu	Cr	Ni	Nb	V	Mo	Ti	Al	Fe
0.027	1.51	0.13	0.014	0.002	0.28	0.27	0.16	0.09	0.004	0.004	0.011	0.035	Bal.

Some of the most important mechanical properties for the API X70 pipeline steel are shown in Table II.

Table II. Typical mechanical properties of the API X70 pipeline steel.

YS (MPa)	UTS (MPa)	E (GPa)	EL (%)	HV
578	660	245	22	221

Ecorr vs. time

Figure 1 shows the corrosion potential as a function of the time of X70 pipeline steel immersed in seawater at different rotation speed. In this figure is possible to observe that the E_{corr} trend to electropositive values (cathodic tendency) as the rotation speed increased. This behaviour is possible to attribute to the fact that, when the rotation speed increased, the turbulent flow facilitates the transport of the cathodic reagent (dissolved oxygen) from the bulk to metallic surface. On the other hand, the anodic reaction is accelerated (metal dissolution) and the ions Fe^{+2} are released; where some of this ions move to the electrolyte and another ions react with oxygen forming oxides that they are adsorbed on surface of the metallic samples. According Nernst equation, the increment of concentration of the ions Fe^{+2} increase the potential [13].

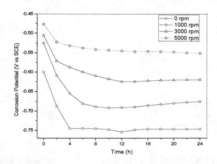

Figure 1. E_{corr} versus time of X70 steel immersed in seawater at different rotation speed.

Qualitative analysis of the EIS spectra

Figure 2 shows the Nyquist diagram with EIS spectra of the X70 steel immersed in seawater at different rotation rate and static conditions. The EIS spectra were analyzed at two different exposure times, 0 and 24 h.

The EIS spectra shows that at the beginning (0h) and at the end (24 h) of the test the charge transfer resistance (R_{ct}) decreased significantly when the turbulent flow (1000, 3000 and 5000 rpm) influenced the corrosion process. It is to mean, the CR values of the X70 steel at turbulent flow conditions are bigger than at static conditions.

Figure 2b shows that at 24 h of the exposure time, the R_{ct} corresponding to static conditions maintain a similar behavior that it presented at the beginning of the test, it is, R_{ct} value is biggest. The behavior presented at this condition is attributed to the fact that a corrosion products film was formed on surface of the steel sample and it should be restricting the charge transfer process and the corresponding reaction of the oxygen reduction causing low values of corrosion rate. In case of the turbulent flow conditions, the R_{ct} decreased as the flow speed increased, this fact indicates that the CR increased as flow speed also increased. This behavior is attributed to the fact that the turbulent flow facilitates the transport of the cathodic reagent (dissolved oxygen) from the bulk to metallic surface and mainly to the effects of the electrolyte movement on the corrosion products film formed on surface of the metallic sample, wall shear stress.

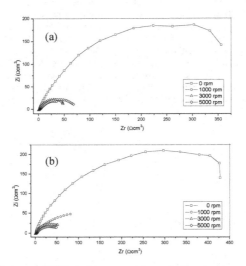

Figure 2. Nyquist diagram of X70 steel immersed in synthetic seawater under different rotation speed and at (a) 0 h and (b) 24 h.

Quantitative analysis of the EIS spectra by equivalent electric circuit (EEC)

According the EIS spectra presented in figure 2, the corrosion process can be analyzed by a simple model of Randles [12]. Figure 3 shows the EEC used and its physic explanation, where R_s correspond to electrolyte resistance, C_{dl} correspond to double layer electrochemical capacitance, R_{ct} is the charge transfer resistance, CR is the corrosion rate, WE is the working electrode, and the RE is the reference electrode. It is important to point out that the EIS spectra shown a depressed below the real axis and this depressed can be explained assuming that the time constant of the process is really a distribution of different time constants, where they are related with the heterogeneity of surface and they could be considered as a collection of the small adjacent electrodes and, each one has its own R_{ct} and C_{dl} and for that reason its time constant [14]. According to EIS spectra depressed, the C_{dl} was substituted by a constant phase element (CPE_{dl}) that it simulates a no ideal capacitor [15]. It is important to point out that the real capacitance was calculated by the CPE_{dl} [16] according to equation 1.

$$C_{dl} = Y_0(\omega_{max})^{n-1} \qquad (1)$$

Where Y_0 y n are constant parameter of the CPE_{dl} and ω_{max} is the maximum frequency.

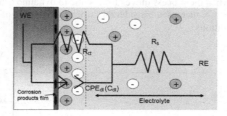

Figure 3. EEC used in the fit of the EIS spectra and its physic explanation [17].

Some researcher [18] indicate that the corrosion products film formed on surface of the metallic samples in the corrosion of the low carbon steel is the hydrated ferric oxide, where, it is not protected and not adherent; for that reason the influence of this film in the CR of the steel is minimum and it is not appear in the EIS spectra (second time constant).

Influence of the wall shear stress and the film thickness on CR
The wall shear stress (τ_w) was measured between the metallic surface (or the corrosion products film absorbed on surface of the steel) and the electrolyte [1]. In a RCE the τ_w is calculated by the equation 2.

$$\tau_w = 0.079 Re^{-0.3}\rho u^2 \qquad (2)$$

Where: Re is the Reynold number, u is the peripheral speed and ρ is the seawater density.
In order to calculate the film thickness, the equation 3 was used [19]. This equation indicates that the capacitance obtained by EEC is the sum of the film capacitance (C_{film}) and the double layer capacitance (C_{dcl}) according to equation 3.

$$C_{film} = C_t - C_{dl1} \qquad (3)$$

Where: C_t is the capacitance obtained by EEC and this was obtained by the literature [20] and its values oscillate between 50-100 $\Box$F/cm^2. The film thickness was obtained by equation 4:

$$\delta_{film} = \frac{\varepsilon \varepsilon_0}{C_{film}} \tag{4}$$

Where ε is the dielectric constant of the corrosion products film, ε_0 is the vacuum permittivity.

The relation between τ_w, CR and δ_{film} is showed in figure 4. This figure shows that τ_w limited the growth of the film thickness, for that reason, la CR increased as the rotation rate also increased, however, in turbulent flow conditions, in the rotation speed from 1000 to 5000 rpm, the δ_{film} was affected by the τ_w, this behaviour should be attributed to the fact that the δ_{film} reach a value of limit thickness. In addition, figure 4 also shows that CR is really affected by the turbulent flow conditions when the process moves from static to dynamic (turbulent flow) conditions, it is to mean, the CR increased considerably when the regimen of the fluid change from static to dynamic conditions.

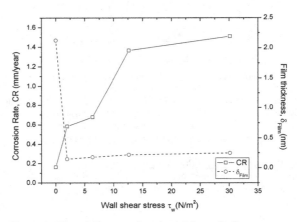

Figure 4. CR and δ_{film} as a function of the wall shear stress

Superficial analysis

Figure 5 shows the micrograph of the X70 steel after it was immersed in synthetic seawater during 24 h at static conditions and different rotation speeds. In this figure is possible to observe that at static and turbulent flow conditions, the morphology of the corrosion process was a localized type. It is important to point out that a 5000 rpm the micrograph show a groove or shallow pit typical of the erosion corrosion.

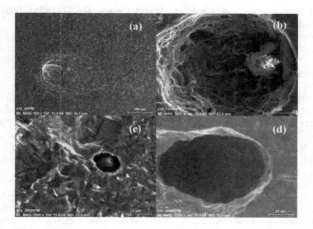

Figure 5. Micrograph of X70 steel immersed in seawater at different rotation speed. (a) 0 rpm, (b) 1000 rpm, (c) 3000 rpm, y (d) 5000 rpm

CONCLUSIONS

According to electrochemical study using the electrochemical impedance spectroscopy technique, the corrosion process of the X70 immersed in synthetic seawater at different rotation rate is possible to conclude that the turbulent flow affected directly the corrosion rate because the CR increased as the flow speed (rotation speed) also increased; this behavior is attributed to the diminishing of the corrosion products film thickness due to the wall shear stress caused by the movement of the fluid. In addition, the fluid movement should facilitate the transfer of the oxygen from the bulk to surface of the metallic sample, in turn; this mass transfer could be accelerating the cathodic reaction. It is important to point out that the corrosion products film formed on surface of the metallic sample is porous and low adhesiveness for that reason it does not appear in the EIS spectra. The morphology of the corrosion process is a localized corrosion form.

REFERENCES

1. H. Edmar Schulz, A.L. Andrade Simões, R. Jahara Lobosco, *Hydrodynamics Optimizing Methods and Tools*, 364, In Tech Rijeka (2011)
2. S.L. Asher, PhD Thesis, Georgia Institute of Technology (2007).
3. B. Poulson, *J. Appl. Electrochem.*, **24**, 1 (1994).
4. D.C. Silverman, *Corrosion.* **40**, 220 (1984).
5. D.C. Silverman, NACE-Corrosion/90, Paper No. 13, USA, (1990).

6. G. Kear, B. D. Barker, K. R. Stokes and F. C. Walsh, *Corros. Sci.* **47**, 1694 (2005).
7. J.L. Mora Mendoza, S. Turgoose, *Corros. Sci.*, **44**, 1223 (2002).
8. S. Nesic, J. Bienkowski, K. Bremhorst, K.S. Yang, *Corrosion*, **56**, 1005 (2000).
9. R. Galván Martínez, R. Orozco Cruz, R. Torres Sánchez, E.A. Martínez, *Mater. Corros.*, **61**, 872 (2010).
10. G. Kear, K. Bremhorst, S. Coles, S.H. Huáng, *Corros. Sci.*, **50**, 1789 (2008).
11. D.R. Gabe, F.C. Walsh, *J. Appl. Electrochem.*, **13**, 3 (1983).
12. J. Genescá Llongueras, *Técnicas Electroquímicas para el Control y Estudio de la Corrosión*, 56, (Programa Universitario de Ciencia e Ingeniería de Materiales, UNAM, México, 2002).
13. E. Otero Huerta, *Corrosión y degradación de materiales*, 67, (Síntesis, Madrid, 2001).
14. E. Barsoukov, J.R. Macdonald, *Impedance Spectroscopy: Theory, Experiment, and Applications*, 2nd ed., 2 (John Wiley & Sons, New Jersey, 2005).
15. A. Hernández Espejel, M.A Domínguez Crespo, R. Cabrera Sierra, C. Rodríguez Meneses, and E. M. Arce Estrada, *Corros. Sci.*, **52**, 2258 (2010).
16. C.H. Hsu, F. Mansfield, *Corrosion*, **57**, 747 (2001).
17. E.C. Corredor, E. Vera, C.A. Ortiz, J.E. Alfonso, *Revista Colombiana de Física*, **38**, 21 (2006).
18. A. Groysman, *Corrosion for Everybody*, 32 (Springer, New York, 2010).
19. R. Galván Martínez, PhD Thesis, UNAM, 2004.
20. F. Mansfeld, H. Shih, H. Greene, C.H. Tsai, *ASTM Special Technical Publication*, **1188**, 37 (1993).

Mater. Res. Soc. Symp. Proc. Vol. 1766 © 2015 Materials Research Society
DOI: 10.1557/opl.2015.415

Electrochemical Study of 1018 Steel Exposed to Different Soils from South of México

L. M. Quej-Ake[*], A. Contreras

Instituto Mexicano del Petróleo, Eje Central Lázaro Cárdenas Norte No.152, Col. San Bartolo
Atepehuacan, Del. Gustavo A. Madero, C.P. 07730, México.
[*] E-mail: lquej@imp.mx

ABSTRACT

Physicochemical effect on the corrosion process of AISI 1018 steel exposed to five type of soils from South of México at different moisture content using electrochemical impedance spectroscopy (EIS) and potentiodynamic polarization curves was studied. Two soils were collected in the state of Veracruz (clay of high plasticity and silt) and three soils from the state of Campeche (sand, clay and clay-silt). Moisture values were determined by addition of 0, 20, 40 and 60 ml of deionized water in a volume of 125 cm^3 of each soil. The corrosion behavior of uncoated and coated steel with a viscoelastic polymer was analyzed. Effect of damage on the coating when the steel is exposed to corrosive soils was studied. EIS evaluations indicate that 1018 steel without coating is more susceptible to corrosion in the clay at the maximum moisture content (39.7 wt. %). However, for sand the more corrosive moisture belong to 12.8 wt. %, which is not the maximum moisture, which is agree with the lower polarization resistance (52.21 $\Omega.cm^2$). Potentiodynamic polarization curves suggested that uncoated steel exposed to clay-silt from state of Campeche exhibited the higher corrosion rate (0.698 mm/year) at 53.1 wt. % moisture. Meanwhile, in the coated steel with induced damage, the higher corrosion rate was obtained in the clay (0.0018 mm/year) at 34.2 wt. % moisture. 1018 steel coated with induced damage exposed to clay displayed the higher E_{corr} values, which means that clay is more susceptible to overprotection as consequence of any change in the voltages originated by moisture content.

INTRODUCTION

The concern with the environment is very important and a better understanding of the soil as a corrosive agent becomes necessary to optimize the cathodic protection systems in the buried pipelines. The corrosiveness soil and the environmental conditions are the main causes of external corrosion and stress corrosion cracking (SCC) of the buried pipe line [1-3]. Another cause of these damages is due to coating failure (physical or mechanical defect) and inefficient cathodic protection (low voltages) or over-voltages in the cathodic protection, which can reduce the structural integrity of buried pipelines. The soil corrosivity can be determined by physicochemical properties like soil type, moisture content, soil resistivity, soluble salts, pH, redox potential and the role of microorganism in the soil. Moreover, microstructure soil properties depend of soil particle size distribution, organic content, mineralogical composition, structure and moisture content, that all of them can change inclusive at short distances

[4].Deterioration in buried pipelines mainly in external surface occurs due to formation of corrosion cells between the metallic structure and the soil type [5]. The steel damage is determined by the change of environment like pluvial and dry season. Nowadays, the studies on steel corrosion in soils and SCC have become of interest in the oil industry and the materials research. Moreover, the corrosion process of pipeline steel using various aqueous solutions representatives of a one type of soil has been developed [6]. In similar way some studies of stress corrosion process of 1018 steel in chloride solutions were carried out [7]. They reported that susceptibility to SCC of the 1018 steel according to different pH and temperatures, follow a linear diffusion behavior. The most likely mechanism for the cracking susceptibility of 1018 steel in the chloride and acid sour solutions seems to be assisted by hydrogen evolution and anodic dissolution which sometimes were called corrosion by hydrogen embrittlement [8]. One study about the effect of hydrogen on fracture characteristics of AISI 1018 steel was carried out [9]. This study indicated that the presence of hydrogen in the steel changes the fracture mode, passing from the microvoid coalescence to quasi-cleavage, accelerating the fracture process. The nonmetallic inclusions in the steel, act as trap sites of hydrogen. When a critical concentration of hydrogen is reached the crack propagates, following a transgranular path due to the embrittlement of the steel. On the other hand, studies associated with the effect of temperature on corrosion process of carbon steel in contact with saline soils and solutions representative of soils suggest that the effect of depth and type of soil are relevant factors to activate the oxidation of steel [10-13]. Due to the soils diversity that exist in México, this work is proposed to use the electrochemical impedance spectroscopy and polarization curves to study the corrosion behavior of AISI 1018 steel with and without coating in contact with five types of Mexican soils with different moisture content. Two soils were collected in the state of Veracruz and three soils from the state of Campeche. An understanding of how coatings fail and the consequences in terms of conditions developed at the steel surface in various soil environments becomes a key issue of this work.

EXPERIMENTAL

Steels used

AISI 1018 commercial steel in form of plates with 2 x 2 x 0.5 cm was used in this study. Table I shows the chemical composition of this steel. Some of the most important mechanical properties for the commercial 1018 steel are shown in Table II.

Table I. Chemical composition of 1018 steel (wt. %).

C	Mn	Si	P	S	Fe
0.18	0.75	0.15	0.03	0.04	Bal

Table II. Typical mechanical properties of AISI 1018 steel.

YS (MPa)	UTS (MPa)	E (GPa)	EL (%)	HV
385	487	190	27	197

Figure 1 shows the microstructure of the 1018 steel used. The low carbon steels has a ferrite-pearlite microstructure, in addition to some manganese sulphide (MnS) inclusions that were observed. These inclusions are considered as preferred locations for corrosion initiation sites or propagation of pre-existing defects [14,15].

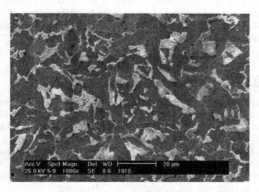

Figure 1. Typical microstructure of 1018 steel.

Soil characterization

Physicochemical and electrical properties of the natural soils collected in South of México were characterized. Two soils were collected in the state of Veracruz and three soils from the state of Campeche. These soils were characterized as was found in the field. After the soils were dried at 60°C for three days, and then the soils were subjected to grinding in order to homogenize the particle size. Subsequently, a constant volume of 125 cm^3 sample of each dried and sieved soil was selected. Then in each soil sample different volumes of deionized water were added (0, 20, 40 and 60 mL), in order to dissolve soluble salts and other ionic species. Subsequently, the physicochemical properties were measured again. Thus, moisture content was determined according to ASTM D-4959 [16]. Redox potential was evaluated according to ASTM G-200 [17] using an Ag/AgCl electrode and pH was measured directly in the soil samples.

Electrochemical evaluations

Electrochemical tests were performed in a cell of acrylic as is shown in Figure 2a. Steel plates with 2 x 2 x 0.5 cm and 1.13 cm² of exposure area from AISI 1018 were used as working electrode (WE); due to the soil is a system with high resistivity and impedance, it is necessary to carry out the IR-drop compensation using three graphite rods that were used like auxiliary electrode (AE) and a saturated calomel electrode (SCE) like reference (RE). The working electrodes were mechanical grinding using emery papers of grade 180, 240, 500 and 600 to obtain a flat, homogeneous and clean 1018 steel surface and then were ultrasonically cleaned with acetone. The steel exposed to the five soils were evaluated by EIS and potentiodynamic polarization curves. All electrochemical experiments were carried out in a Potentiostat-Galvanostat model PGSTAT30 controlled by a PC supported by the General Purpose Electrochemical System (GPES) and Frequency Response Analyzer (FRA) software provided by AUTOLAB. The impedance spectra were performed in the frequency range of 10 kHz to 10 mHz with an amplitude of 10 mV, recording 7 points per frequency decade. Polarization curves were recorded at 0.001 Vs⁻¹ and the potential range used ± 0.3 V referred to E_{corr}.

Cathodic protection of induced defects on the coating

In order to analyze the behavior of coating with a defect, a mechanical damage around 3 mm in diameter was made in the coating (0.0706 cm²) and cathodic potentials of -850, -950, -1200, -1500 and 1700 mV were applied on 1018 steel for a period of 30 days. For the cathodic protection tests, the negative terminal from a power source was connected to the working electrode considered as the cathode; thereafter the positive pole of the power source was connected to a graphite rod considered as sacrificial anode as shown in Figure 2b. Subsequently, the corrosion potential values between steel electrodes and the soil were determined using Cu/CuSO₄ as the reference electrode and a Potentiostat-Galvanostat.

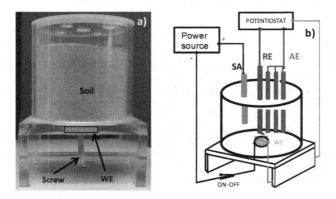

Figure 2. Experimental set-up used to carry out the electrochemical tests at room temperature.

RESULTS AND DISCUSSION

Physicochemical evaluations of the soils

Figure 3 shows images of the five soils collected in South of Mexico. Two soils were collected in the state of Veracruz (clay of high plasticity and silt) and three soils from the state of Campeche (sand, clay and clay-silt). As can be noted, the soils present colorations and particle size different, which could be associated with changes in the physicochemical properties.

Figure 3. Images of the physical view of the soils studied.

Table III show the physicochemical properties of the soils measured in field directly (0 % water added) and the properties after 20, 40 and 60 mL of water added.

Table III. Physicochemical properties of the soils measured in field (0% water added) and measured in lab (adding 20, 40 and 60 mL of water).

Soil	Location	Water added (mL)	Moisture (wt.%)	pH	Redox Potential (mV vs. Ag/AgCl)
Clay of high plasticity (CHP-V)	Near to Zapoapita town (Veracruz)	0	7.02	6.5	-253
		20	12.22	6.65	-356
		40	32.11	6.69	-357
		60	40.75	6.75	-366
Silt (Silt-V)	Near to Coatzacoalcos river (Veracruz)	0	3.43	6.0	-9.35
		20	21.11	6.11	-288
		40	29.45	5.62	-375
		60	36.49	5.35	-399
Sand (Sand-C)	Near to the coast (Campeche)	0	0.08	6.4	-3.65
		20	9.82	6.79	-350
		40	12.80	7.05	-393
		60	24.92	7.27	-409
Clay (Clay-C)	Near to highway Campeche to Mérida (Campeche)	0	3.63	6.0	-50
		20	20.35	6.22	-411
		40	27.85	7.06	-332
		60	39.71	7.67	-297
Clay-Silt (CS-C)	Near to highway Campeche to Mérida (Campeche)	0	5.48	6.8	-120
		20	21.04	7.21	-319
		40	27.42	7.35	-308
		60	53.12	7.75	-301

It is observed that most of the soils from the state of Campeche have near neutral pH; soils obtained from Veracruz have pH values with acid trend. The clay from Campeche (Clay-C) provides the more negative redox potential value (-411 mV) at 20.35wt.% moisture. This indicates that the clay acts as reducing soil that could be affect metal oxidation. The moisture, pH and redox potential presented in Table III, show the complexity of physicochemical phenomena that provide the soils when they have different water content, which indicates that this variation significantly affect the corrosion process of steels.

Electrochemical impedance spectroscopy (EIS)

Figure 4a shows the Nyquist plots for 1018 steel in contact with the five types of soils evaluated at the maximum water content added (60 mL) and 1 h of exposition. According to the magnitude of the impedance values, it is observed that Clay-Silt from Campeche (CS-C) is the less corrosive soil, while the clay from Campeche (Clay-C) is the most corrosive soil.

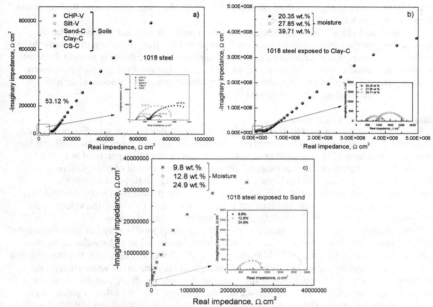

Figure 4. Nyquist plots obtained for 1018 steel a) exposed to the five soils at the maximum moisture content (adding 60 mL of water) and 1 h of exposition, b) exposed to Clay-C, c) exposed to Sand-C with different moisture content.

The corrosive effect of the Clay-C can be attributed by the lower redox potential value at its maximum moisture content (-297 mV). This indicates that the Clay-C also acts as oxidizing soil that could be affecting the reduction process on metal surface. On the other hand, the moisture evolution provides a greater dispersion of the ions contained in the electrolyte. There is a relationship between solute and solvent such that the high concentration of the ionic conductor favors the process of corrosion of steel.

Effect of the moisture content

Figure 4b shows the real and imaginary impedance values for 1018 exposed to Clay-C at three moisture content after 1 h of exposition. In this figure can be observed that soil with highest moisture (39.71 wt.%) activates the corrosive effect on 1018 steel. While at low moisture evolution (20.35 wt. %) the same steel manifested corrosion resistance.

Figure 4c shows the impedance values for 1018 exposed to Sand-C at three moisture content. In this case it can be observed that sand with intermediate moisture (12.8 wt. %) was more corrosive. While at high moisture content (24.9 wt. %) the same steel manifested corrosion resistance as well as 9.8 wt. % moisture.

The sands and clays behave as ionic conductors in presence of water, but when the water begins to evaporate (low moisture), this soils will not permit the flow of current efficiently. In underground pipelines, this phenomenon could be dangerous because dry clay or sand does not allow adequate cathodic protection, however, when the soils changes from dry to wet, it could cause over-potentials that would harm the coating (disbonding) and can produce chemical reactions that induce hydrogen embrittlement [3,9,14].

Qualitative analysis of EIS spectra

In order to carry out a quantitative analysis of the impedance spectra, an adjustment of the experimental data was performed by means of equivalent circuits and the program nonlinear least squares fit of Boukamp [18]. Due to the existence in the majority of the spectra of at least a semicircle, for adjustment of the experimental data an equivalent circuit of a single time constant R_s ($R_p C_{EDL}$) was used in order to study the corrosion process, where R_s is the resistance associated with the conductive properties of soil used, R_p is the polarization resistance and C_{EDL} is the capacitance associated with the electrical double layer. Table IV shows the values of the electric parameters obtained by the best fit of the experimental data obtained.

Table IV shows that lower resistance of the soil (R_s) correspond to clay-silt from Campeche (CS-C), indicating that low resistance is associated with an increased of soil conductivity, thus this type of soil behaves as ionic conductor, corrosive at 27.42 wt.% of moisture. For the different soils, moisture change according to soil type, indicating that the addition of water changes the ionic strength of the soil. The polarization resistance is associated with the transfer of ionic and electronic charge generated in the steel surface and the solution. The low value of polarization resistance indicates a greater charge transfer activity in the electric double layer [19,20] thus the low value for R_p is indirectly associated with an increased in the steel corrosion process. Thus, it is possible to identify that the sand is the most corrosive soil with $R_p = 52.21$ Ω cm^2 at a moisture of 12.8% follow by Clay-C with $R_p = 874.07$ Ω cm^2 at a moisture of 39.71%.

When the 1018 steel is exposed to Silt-V, Clay-C and CS-C the major corrosiveness belong to the soil with the maximum moisture content. By the contrary, when the 1018 steel is exposed to CHP-V and Sand-C, the major corrosiveness is not obtained at the major moisture content.

The capacitance values shown in Table IV for the properties of the electric double layer were calculated using equation 1 [21]:

$$C = \frac{(Y_o R)^{\frac{1}{n}}}{R}$$

(1)

Table IV. Resistance and capacitance for 1018 steel exposed to soils at different moisture content after 1 h.

Soil	Moisture content (Wt. %)	R_s ($\Omega.cm^2$)	R_p ($\Omega.cm^2$)	C_{EDL} ($\mu F.cm^{-2}$)
CHP-V	12.22	1632	1634	241.3
	32.11	1058	2773	204.2
	40.75	1265	3572	138.4
Silt-V	21.11	5604	5733	9.6
	29.45	2250	2666	173.6
	36.49	1276	1233	357
Sand-C	9.820	113030	81.01 M	80.756 pF
	12.80	358.87	52.21	558.4
	24.92	402.49	3182	39.8
Clay-C	20.35	2.03 M	38.73 M	6.65 pF
	27.85	868.76	1226	76.65
	39.71	400.27	874.07	856.69
CS-C	21.04	742490	655.96 M	1.67 nF
	27.42	292.4	60.33 M	24.83 nF
	53.12	77626	5.78 M	100.01 nF

M=Mega, n=Nano, p=pico

It is known that depending on the value of capacitance, the active area and the spacing between steel surface and electrolyte solution is changed [22]. The capacitance assessment (C) was carried out in function of the area (A), charge (q), the dielectric constant (ε), the disturbance voltage (E) and the spacing between the parallel flat plates (δ) as described in equations 2-3. Where C_i is the capacitance value of an i (μF) system; ε_o is the vacuum dielectric constant (8.85×10^{-14} Fcm^{-1}), ε_i is the dielectric constant of the i system; δ correspond thickness; dq is the change of the charge (Coulomb) dE is the change in potential (Volts).

$$C_i = \left(\frac{\varepsilon_0 \varepsilon_i A_i}{\delta_i} \right)$$

(2)

$$\delta \leq C = \left(\frac{dq}{dE} \right) \geq \delta$$

(3)

This indicates that a greater R_p, a lower contribution of C is generated, indicating the existence of a lower electrochemical activity and large δ in the electrical double layer (EDL) and a suitable electric field oriented in the steel-soil interface, which control the corrosion process.

Potentiodynamic polarization curves

Corrosion in steel-soils interface

Figure 5 shows potentiodynamic polarization curves obtained for 1018 steel exposed to five soils studied at maximum moisture content (adding 60mL of water) and 1 h of exposition. This figure show that corrosion potential obtained in CHP-V and Silt-V are located in the more negative zone, indicating that these soils types have higher tendencies to metal oxidation.

However, due to the higher values of anodic (oxidation branch) and cathodic (branch reduction) current density obtained for the steel exposed to CS-C, suggests that the 1018 steel is subjected to an diffusion-activation process, allowing greater charge transfer and electrons between the ionic conductor (soil) and the metal (electronic conductor).

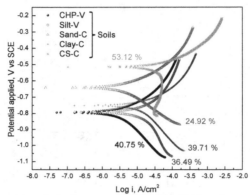

Figure 5. Potentiodynamic polarization curves obtained for 1018 steel exposed to the five soils at the maximum moisture content (adding 60mL of water) and after 1 h.

Table V shows the corrosion rates values obtained from polarization curves for 1018 steel exposed to saturated soils. From these results, it is observed that CS-C is the soil more aggressive for the 1018 steel with a corrosion rate of 0.698 mm/year.

Table V. Corrosion rates obtained from polarization curves for 1018 steel.

Soil	Moisture content (Wt. %)	E_{corr} (mV vs. SCE)	CR (mm/year)
CHP-V	40.75	-794	0.0309
Silt-V	36.49	-791	0.0488
Sand-C	24.92	-641	0.0462
Clay-C	39.71	-751	0.1625
CS-C	53.12	-514	0.6989

Effect of coating damage

It is well known that underground pipe line steel is protected by two forms, one of them is a physical barrier (coating) and the other is electrochemical (galvanic anodes or impressed current). However, is a common find defect with different geometry and morphology on the coating, produced by mechanical damage and deterioration. For that reason it is useful to understand the effect of the damage on the coating when the steel is exposed to a corrosive soil. In this way, were selected the three more corrosive soils (Clay-C, CS-C and Sand-C) to carry out electrochemical evaluations and study the corrosion process on 1018 steel coated with one defect. Figure 6a shows EIS spectra obtained after been exposed 30 days the 1018 steel to the soils.

From this figure it is possible to observe that steel is more susceptible to corrosion when it is exposed to Clay-C. It is important to mention that the electrochemical phenomena shown in these profiles are associated with the metallurgical properties of 1018 steel and the response of the induced defect area. Thus, considering that the defect on the coating has the same area, the EIS changes are due to the presence of these types of soils. The defect in the coating is a localized damage that induced a greater active area and the damage evolution can provoke crevice corrosion.

Figure 6b shows potentiodynamic polarization curves obtained for 1018 steel exposed to three saturated soils from Campeche. The cathodic branches show the contribution of different changes of slopes; after a second slope is observed which is associated to the reduction-diffusion process.

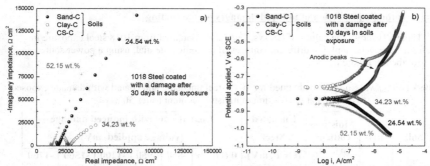

Figure 6. a) Nyquist plots and b) Potentiodynamic polarization curves obtained for 1018 steel coated with an induced damage exposed to the three soils from Campeche at the maximum moisture content (adding 60mL of water after 30 days).

The anodic branch shows a slight decrease in the value of the oxidation current of about 100-120 mV, after these potential values is observed a significant increase in the drain current. The slight decrease in current indicates that the passivation or the formation of corrosion products obstructs the process of mass and charge transfer between the steel-coating-soil; however at lower voltages an anodic peak is observed in each system, promoting the activation (pitting) and dissolution.

Table VI lists physicochemical parameters of the soils from Campeche after 30 days exposure and corrosion rate obtained from polarization curves for 1018 steel coated with an induced defect. It is observed that Clay-C presents the highest corrosion rate for 1018 steel, which is agree with results of Figure 6a and 6b. It is observed that the same coating on the steel acts different depending of soil exposure, it could be that the different soil modifies its adhesion properties changing the exposure area and the corrosion process.

Table VI. Physicochemical parameters of soils and corrosion rate for 1018 steel coated with damage exposed to soils from Campeche at the maximum moisture content (after 30 days).

Soil	Moisture content (Wt. %)	Redox potential (mV vs. Ag/AgCl)	pH	E_{corr} (mV vs. SCE)	CR (mm/year)
Sand-C	24.54	-301	6.3	-827	0.0010
Clay-C	34.23	-265	7.5	-762	0.0018
CS-C	52.15	-299	6.8	-839	0.0005

Effect of cathodic protection of induced damages on the coating

Table VII shows the values of E_{corr} vs. $Cu/CuSO_4$ obtained from 1018 steel exposure in soils from Campeche applying different voltages of cathodic potential, using a power source as was described in Figure 2.

Table VII. E_{corr} vs. $Cu/CuSO_4$ obtained for 1018 steel uncoated and coated with damage exposed to the soils at the maximum moisture content (after 30 days).

Soil	pH	Moisture content (wt. %)	Uncoated Steel (OCP, mV)	Coated steel with induced damage (voltage applied, mV)					
				0 (OCP)	-850	-950	-1200	-1500	-1700
Sand-C	7.61	22.75	-719	-755	-821	-901	-1028	-1085	-1230
Clay-C	7.61	30.11	-807	-835	-849	-938	-1116	-1140	-1292
CS-C	7.34	47.91	-692	-799	-828	-913	-1046	-1122	-1277

Table VII shows that E_{corr} (mV vs. $Cu/CuSO_4$) obtained at open-circuit potential (OCP) for uncoated steel were lower than those obtained in the steel coated in presence of damage. In order to meet the cathodic protection criterion (-850mV), it is necessary to apply -950 mV.

The potential distribution in the three soils through steel changed E_{corr} values toward random values. This effect is caused by osmic drop (IR drop) because each soil had typical resistance (R_s).

On the other hand, E_{corr} values for 1018 steel coated with induced damage exposed to Clay displayed higher values than Sand and CS-C, which means that Clay is more susceptible to overprotection as consequence of any change in the voltages and consequently will provides a higher damage in the coating (disbonding) and steel (hydrogen embrittlement) [23,24]. Studies of coatings with a simulated damage can be useful to predict the corrosion protection and properties of coating degradation and adhesion in pipelines.

CONCLUSIONS

This investigation was focused on physicochemical effect of different soils from South of México with different moisture content on the corrosion process of AISI 1018 steel. The soils evaluated have different water retention and each soil has a typical moisture content in which the corrosion process is more active. The greater corrosion rate, do not belong necessarily in saturated soils. From polarization curves for 1018 steel exposed to saturated soils, it is observed that CS-C is the more corrosive soil for 1018 steel with a corrosion rate of 0.698 mm/year; meanwhile, for 1018 steel coated with damage the higher corrosion rate (0.0018mm/year) was obtained in Clay. The results suggest the use of EIS and polarization curves in the cathodic protection study is a really important tool for the analysis of the corrosion process in underground pipelines at different stages of season of the year (dry and rainy), due to the cathodic potential variations in different types of soils and different moisture contents.

REFERENCES

1. L.M. Quej-Ake, R. Galván Martínez, A. Contreras, *Materials Science Forum*, **755**, 153 (2013).
2. T. Haruna, T. Shibata, R. Toyota, *Corrosion Science*, **39**, 1935 (1997).
3. A. Contreras, S.L. Hernández, R. Orozco Cruz, R. Galván Martínez, *Materials and Design,* **35**, 281 (2012).
4. J.L. Alamilla, M.A. Espinosa Medina, E. Sosa, *Corrosion Science*, **51**, 2628 (2009).
5. S.A. Bradford, Practical Handbook of corrosion control in soils, Third printing, Canada, Casti, (2001).
6. A. Benmoussat, M. Hadjel, *The Journal of Corrosion Science and Engineering*, **7**, 1 (2005).
7. M.A. Espinosa Medina, E. Sosa, C. Angeles Chavez, A. Contreras, *Corrosion Engineering, Science and Technology*, **46**, 32 (2011).
8. L. Quej, R. Cabrera, E. Arce, J. Marin, *International Journal Electrochemical Science*, **8**, 924 (2013).
9. G. González, V.J. Cortez, J.G. Ramírez, *Revista Mexicana de Física*, **50**, 60 (2004).
10. X.H. Nie, X.G. Li, C.W. Du, Y.F. Cheng, *J. Applied Electrochemistry,* **39**, 277 (2009).
11. D. Gervasio, I. Song, J.H. Payer, *J. Applied Electrochemistry*, **28**, 979 (1998).
12. J.N. Murray, P.J. Moran, *Corrosion*, **45**, 885 (1989).
13. J.H. Fitzgerald, *Materials Performance, ***49**, 17 (1993).
14. Z. Velázquez, E. Guzman, M.A. Espinosa, A. Contreras, *Materials Research Society Symposium Proceedings*, **1242**, 69 (2010).

15. A. Contreras, S.L. Hernández, R. Galvan, *Materials Research Society Symposium Proceedings,* **1275,** 43 (2011).
16. ASTM D-4959, *Standard test method for determination of water (moisture) content of soil by directs heating,* (2007).
17. ASTM G-200, *Standard test method for measurement of oxidation-reduction potential (ORP) of soil,* (2014).
18. B.A. Boukamp, Users Manual Equivalent Circuit, Version 4.51, Faculty of Chemical Technology, University of Twente, Netherlands (1993).
19. E.E. Stansbury, R.A. Buchanan: *Fundamentals of Electrochemical Corrosion,* United States of America: ASM International, first Edition, (2000).
20. P. Marcus, J. Oudar, *Corrosion Mechanisms in Theory and Practice,* New York: Marcel Dekker, Inc., First Ed., (1995).
21. M.A. Pech Canul, L.P. Chi Canul, *Corrosion,* **55,** 948 (1999).
22. J.R. Macdonald, *Impedance Spectroscopy,* United States of America, John Wiley and Sons, First edition (1987).
23. M. Yan, J. Wang, E. Han, W. Ke, *Corrosion Science,* **50,** 1331 (2008).
24. H. Bi, J. Sykes, *Corrosion Science,* **53,** 3416 (2011).

Mater. Res. Soc. Symp. Proc. Vol. 1766 © 2015 Materials Research Society
DOI: 10.1557/opl.2015.416

The Role of Calcareous Soils in SCC of X52 Pipeline Steel

A. Contreras[1*], L. M. Quej-Aké[1], C. R. Lizárraga[2], T. Pérez[2]

[1] Instituto Mexicano del Petróleo, Eje Central Lázaro Cárdenas Norte 152, Col. San Bartolo
Atepehuacan, C.P. 07730, México D.F.
*Contacting author email: acontrer@imp.mx

[2] Universidad Autónoma de Campeche, Campeche, México, ave. Agustín Melgar s/n, Col
Buenavista, P.O. Box 24039, Campeche, México.

ABSTRACT

Susceptibility to external stress corrosion cracking (ESCC) of API X52 pipeline steel in calcareous soil containing different moisture content has been investigated using slow strain rate tests (SSRT). This type of soil is common of the state of Campeche Mexico and has a pH around 8. The results indicate that X52 pipeline steel was susceptible to external SCC only in the saturated calcareous soil, showing some micro cracks in the gage section of the SSR specimen. It was observed that some micro cracks were found at the bottom of micro-pits. Which indicate that first develop a pit and this evolved with time and micro-strain like a crack. Few micro cracks were observed as initiation of SCC close to surface failure. The mechanism of SCC may be influenced by formation and rupture film of carbonates.

INTRODUCTION

The investigation and studies on SCC originated by the soils in the current days have become of interest in the oil industry in Mexico due to several cases of SCC have been detected through smart pigs before to occur a failure in the pipeline [1]. SCC is dependant of the time and can develop in pipelines under normal operating condition when a coating disbonded and ground water comes into contact with the outside surface of the pipe.

SCC is a type of environmentally assisted cracking that occurs when tensile stresses (mechanical factor) are applied to a susceptible material (metallurgical factor) which is exposed to corrosive environment (electrochemical factor). SCC can develop in pipelines under normal operating condition when a coating disbonded and ground water comes into contact with the outside surface of the pipe.

SCC processes involve complex interactions between metallurgy, operational and residual stresses, external soil environment, and the electrolyte chemistry beneath disbonded coatings. During the years has been observed that some pipelines suffer cracks when they are subject to stresses (residual or operational) in a corrosive soil. SCC in pipelines has been found to occur in a range of soils, in a range of diameters, thicknesses, grades, manufacturing processes and coatings [2].

SCC produced by the soils must taking account many factors like the moisture content, pH, temperature, concentration of dissolved ions, carbonates content, redox potential, resistivity, level of cathodic protection, operational and residuals stresses and time of electrolyte in contact with the steel.

All pipe steels used for oil and gas transmission pipelines are susceptible to SCC [3]. The lower SCC susceptibility observed on some of the most recent pipes grades (X80 and X100) might be due to metallurgical factors, better coating systems, or improvement in the cathodic protection systems [4-8].

SCC may be associated with intergranular or transgranular paths through the metal and in some cases with a mixture of both modes [9-12]. High pH SCC generally produces intergranular cracking and occurs only in a relative narrow cathodic potential range and at pH greater than 9. For near neutral pH SCC the fracture is generally transgranular with pH between 6 and 8[3, 13-15]. The cracks in both types of fracture occur usually in the outside surface of the pipeline in colonies oriented longitudinally, and they propagate in the direction normal to the applied stress.

SCC with near neutral pH has been observed occurring more often in poorly drained soils where reducing conditions are maintained [16]. These soils contain low concentrations of carbonate-bicarbonate, with the presence of other species, including chloride, sulphate, and nitrates.

Most of the cases of SCC produced by the soils are caused by circumferential stresses related with the presence of bicarbonate-carbonate (HCO_3^- — CO_3^{2-}) in the soil water [14, 17]. It has been established that CO_2 — HCO_3^- — CO_3^{2-} is always present in the soil water, but the importance is determine the content of these carbonates.

Mexico it is a country with a wide variety of climates and environmental conditions; it is possible to find cold, hot, mild, wet, and warm climates; most of the pipelines transporting hydrocarbides are located in the South of Mexico, where is very common to find calcareous soils. This work proposed to study the physicochemical effect of calcareous soil in the SCC susceptibility of X52 pipeline steel.

EXPERIMENTAL

Steel used

One of the most used steels in the oil industry in México is X52 steel, for that reason the material used in this study was API 5L X52 pipeline steel. The chemical composition of the steel is shown in Table I.

Table I. Chemical composition of the API X52 pipeline steel (wt %)

C	Mn	Si	P	S	Cu	Cr	Ni	Nb	V	Ti	Al	Fe
0.08	1.05	0.26	0.019	0.003	0.019	0.02	0.02	0.041	0.054	0.002	0.038	Bal.

The X52 steel microstructure consisted of fine pearlite and ferrite with a grain size around 10–20 microns. The X52 steel has yield strength of 390 MPa, ultimate tensile strength of 495 MPa and 24% elongation.

Soil used

Calcareous soil obtained from the south of México was used in this study. The soil samples were obtained from Campeche State. Much of the pipelines transporting hydrocarbides are in the south of México, which was the reason to focus the study on these soils. The calcareous soil used has a pH around 8. SCC occurs in the local environment deriving from the surrounding soil groundwater. For that reason seasonal variation in soil parameters is very important, because all the parameter related to SCC are modified. Thus, calcareous soil with 25, 50 and 100% moisture content was studied.

Specimens for SSRT

The specimens used in this study were obtained from API 5L X52 pipeline steel and were machined according to NACE TM 0198 [18]. The geometry and dimensions of the specimens for SCC experiments is shown in Fig. 1. The length direction of the sample was parallel to the circumferential direction of the pipeline steel in order to ensure that the subsequent crack growth was in the longitudinal direction of the pipe as is typically observed in field.

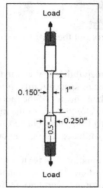

Figure 1. Specimen used to perform the SSRT.

Slow strain rate tests (SSRT)

In the last years SSRT have been used to evaluate SCC in metallic materials. The SSR test has emerged as a relatively quick, simple method that can be used for the evaluation of corrosion resistant alloys (CRA) for resistance to a variety of environmental cracking phenomena [18]. The principal effect of the constant extension rate, in combination with corrosive environment, is to accelerate the initiation of cracking in susceptible material.

SSRT test is a very effective technique for screening cracking environments of underground pipelines. SSRT provides useful information on SCC susceptibility of the materials in any corrosive environments, in addition to a relatively short experimental time to evaluate SCC susceptibility [19, 20].

Experimental set-up

A mobile constant extension rate tests machine (MCERT) with load capacity of 44 kN and total extension of 50 mm as is shown in Fig. 2a was used to perform the SSRT at 1×10^{-6} s^{-1} strain rate. The SSRT were performed in air (Fig. 2b) and in calcareous soil (Fig. 2c) with different moisture content. To perform the SSR tests in the soil, a 500 mL glass autoclave as is shown in Fig. 2c was used. The soil samples used for experiments were prepared at controlled water content corresponding to 25, 50% and 100% of the saturation level.

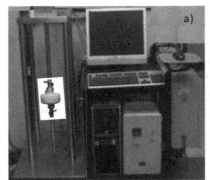

Figure 2. a) MCERT machine used to carry out the SSR tests, b) in air, c) in calcareous soil.

Stress corrosion cracking assessment

After carried out the SSRT, SCC susceptibility was evaluated according to NACE TM-0198 and ASTM G-129[18, 21]. The degree of susceptibility to SCC is generally assessed through observation of the differences in the behavior of the mechanical properties (including reduction in area, time to failure and plastic elongation) of the material in tests conducted in a specific environment (in this case the soil) from that obtained from tests conducted in the controlled environment (air). More details of the equations and criteria used in the assessment were reported elsewhere [18, 22-25].

Complementary metallographic examination was performed to establish whether or not there is SCC on the samples. The presence of cracks was evaluated on the longitudinal section of the gage. Overalls when there is some uncertainty in the assessment of SCC susceptibility evaluating the mechanical properties; and to differentiate in the cases of corrosion is present instead of SCC.

RESULTS AND DISCUSSION

Steel microstructure

Fig. 3 shows the microstructure obtained by optical microscopy of the API X52 steel used in this study. The structure consists of dark areas (perlite) and the light areas formed by ferrite. Low carbon steels generally have a ferrite-pearlite structure containing little pearlite.

Figure 3. Typical microstructure of the API X52 steel.

Soil analysis

Table II shows the chemical composition of soil used in the study. The main constituent is calcium, which is observed forming some carbonates, which were detected through X-ray diffraction. Calcite ($CaCO_3$), Coesite (SiO_2) and magnetite (Fe_3O_4) were detected. High sodium or potassium levels might promote development of concentrated carbonate/bicarbonate solutions under disbonded coatings in pipelines.

Table II. Chemical composition of the soil used.

Metal	Content
Ca	39000 ppm
Mg	766.1 ppm
Si	405.3 ppm
Al	258.7 ppm
Fe	208.8 ppm
K	53.20 ppm
Na	38.45 ppm

The carbonate dissolution will bring the soil from neutral to alkaline pH range. Exposure of this saturated solution to steel surfaces make alkaline by electrochemical reactions induced by cathodic protection system, which will precipitate hard white carbonate scales (e.g., calcite $CaCO_3$ or dolomite $CaMg(CO_3)_2$) on the metal surface. These scales are strongly adhered and form an impermeable protective layer. However, theses layers are brittle and when they break gives origin to dissolution and cracks in the metal surface. Most of the SCC cases in pipelines with near neutral pH, CO_3^{2-} have been found in the metal-coating interface [11, 13, and 26].

High pH SCC has been observed in solutions with various ratios of sodium carbonate to sodium bicarbonate ranging from almost pure sodium bicarbonate to almost pure sodium carbonate [27]. Those ratios correspond to a pH range from about 8 to 10. SCC is most severe in highly concentrated solutions. Some others physicochemical parameters related to corrosion as well as parameters related to SCC susceptibility (pH, moisture, carbonates) were obtained. These parameters are shown in Table III.

Table III. Physicochemical parameters of the soils relate to corrosivity and SCC.

Resistivity (Ohm.cm)	pH	Redox potential (mV)	Organic Matter	Moisture content	CO_3^{-2} (ppm)	HCO_3^- (ppm)
13417	8.1	-335	3.23 %	22.52 %	120	597.8

Through ionic chromatography, the main ions were analyzed in the soils samples. The results are shown in Table IV. Cl and SO_4 even in low percentages may promote soil acidity. The Cl ions are particularly significant in the corrosion process, since they inhibit the formation of passive films on the steel, and promote the pitting corrosion due to the high ionic conductivity.

Table IV. Anions content in the soil studied.

Ion (ppm)	F^-	Cl^-	NO^{-2}	NO^{-3}	SO_4^{-2}
Calcareous	0.49	0.52	2.21	1.56	1.00

Previous studies [25, 28-30] with X52 pipeline steel were performed using a simulated ground water solution (called NS4) as the corrosive environment. NS4 synthetic solution has been widely used to simulate the soil solution in the study of near neutral pH-SCC behavior. However, in the present work the soils were placed directly in contact with the steel samples, using different moisture content in order to simulate the seasons of the year (dry and rainy).

The soils containing carbonates have been related to improve the probability to develop SCC [3, 7-15, 26, 27]. Calcareous deposits of surface film comprised primarily of $CaCO_3$ and Mg $(OH)_2$ that precipitate on cathodic surfaces like a consequence of increased pH near the metal-electrode interface in association with the following reactions:

$$CO_2 + H_2O \rightarrow H_2CO_3 \quad (1)$$

$$H_2CO_3 \rightarrow H^+ + HCO_3^- \quad (2)$$

$$HCO_3^- \rightarrow H^+ + HCO_3^{2-} \quad (3)$$

Calcareous and magnesium deposits would form on pipe surface at the opening where the soil solution contained sufficient amount of Ca^{2+} and Mg^{2+} ions according to the following reactions:

$$Mg^{2+} + 2OH^- \rightarrow Mg(OH)_2 \quad (4)$$

$$Ca^{2+} + CO_3^- \rightarrow CaCO^{3-} \quad (5)$$

The importance of calcareous deposits to the effective and efficient operation of underground pipelines in the cathodic protection systems is not fully understood. Jack *et* al [31] studied the different electrolytes associated with high and near neutral pH SCC in the field and discuss the processes responsible for creation of solutions responsible to the cracking in pipeline steel. The water samples have contained some bicarbonate ions plus lesser amounts of carbonate, chloride, and sulfate. The major cations are sodium, calcium, potassium, and magnesium. Soils near SCC sites have been found to contain 4 to 23 percent CO_2 [11, 32].

Slow strain rate tensile tests

Fig. 4 shows the stress-strain curves obtained from the SSRT carried out in air and in the calcareous soil with different moisture content at room temperature. According to this figure it is clear that strain decreases as the moisture content increase. This can be related to a damage produce by the corrosive solution in the specimen. It is also suggested that specimens suffer hydrogen embrittlement.

Galvanic corrosion, in the form of a stress cell (regions with different stresses), can occur along the specimen gage length when the specimen is exposed to long time loading in an aqueous environment. The higher stress and plastic strain locations become anodic to the lower stress locations. This observation suggests it is the stress-strain differential that is important in causing micro-pitting and eventual SCC.

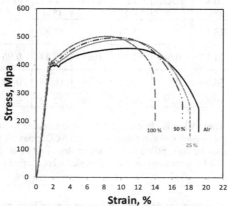

Figure 4. Stress-Strain profiles obtained from the SSRT.

Assessment of SCC susceptibility

Susceptibility to SCC of API X52 steel was evaluated from the results obtained from the SSRT and corroborated with SEM observations. Table V shows a summary of the mechanical properties evaluated, meanwhile Table VI show the SCC assessment results. The SCC susceptibility was expressed in function of reduction in area ratio (RAR), time to failure ratio (TFR) and plastic elongation ratio (PER). These ratios are obtained from comparing the mechanical properties obtained from SSRT in the soils with the mechanical properties obtained in the controlled environment (air). The failure time was between 40 and 55 hours.

Experimental data from final diameter of fracture surface (D_f) to calculate the RRA were obtained from SEM measurements. An average of the measurements was obtained in order to calculate the reduction area ratio. Ratios of RAR, TFR and PER were calculated using the equations of NACE TM198 and ASTM G129 standards [18, 21]. Ratios in the range of 0.8-1.0 normally means high resistance to SCC, whereas low values (i.e. <0.5) indicate high susceptibility to SCC [18-20, 21-25].

Table V. Mechanical properties obtained from SSRT to evaluate SCC.

Condition	Yielding strength (MPa)	Ultimate tensile strength (MPa)	Elongation maximum (mm)	Strain maximum (%)	Plastic elongation (%)
Air	396	460	4.85	19.10	17.82
25%	402	490	4.60	18.11	17.11
50%	409	498	4.36	17.19	16.78
100%	406	502	3.55	13.98	12.72

Table VI. SCC assessment obtained from the SSR tests.

Condition	RA (%)	RAR	TF (h)	TFR	PE (%)	PER
Air	78.59		55.78		17.82	
25%	52.13	0.66	52.97	0.95	17.11	0.96
50%	54.34	0.69	50	0.90	16.78	0.94
100%	59.81	0.76	40.77	0.73	12.72	0.71

According to results of Table VI, it is clear that X52 steel is resistant to SCC, except in the case of samples tested in calcareous soil saturated, where all SCC index were below 0.8, which is indicative that steel may be susceptible to SCC under this conditions. In order to corroborate this assumption, SEM observations of these samples were carried out.

Surface fracture analysis

The de SSRT makes very difficult to determine the exact type of fracture, in most of the samples seems to be a combination of both transgranular and intergranular fracture. For all the samples tested with both soils, it was observed a ductile type of fracture. Ductile fractures are characterized by extensive plastic (permanent) deformation of the material. One mechanism of ductile fracture is known as microvoid coalescence [24, 29, and 30]. The surface fracture showed some microvoids which were originated in some cases by the presence of some inclusions.

Microstructure of steel has been considered like one critical factor affecting stress corrosion initiation and propagation [4, 6-8, 15, 24, 29]. There exist several kinds of inclusions in pipeline steel due to its complex alloy elements. Inclusions are mainly identified as aluminum oxides, calcium oxides, calcium sulfide, and manganese sulfides among others.

Generally, inclusions enriching in Al_2O_3, MnS and SiO_2 exist in pipeline steels. According to Liu et al. [33] the inclusions acts as the crack initiation sites. Whether a crack initiates at an inclusion depends on the inclusions composition and morphology. Liu et al. [54] said that inclusions enriched in Al_2O_3 and SiO_2 are hard and brittle as well as incoherent to the metal matrix.

Longitudinal sections analysis

All the specimens were longitudinally sectioned and polished to SEM observations for inspection of micro-cracking in the gage section. Sometimes it may be necessary to chemically clean the specimen to facilitate adequate inspection. The presence of secondary cracks close to the surface failure is very important to determine if SCC exists. The assessment of SCC was made taking account the results of RAR, TFR and PER, in combination with the results of SEM, observations in the gage section of samples in order to determine if secondary cracks were presents. Fig. 5 shows the SSR specimens after been fractured and before to be longitudinally sectioned.

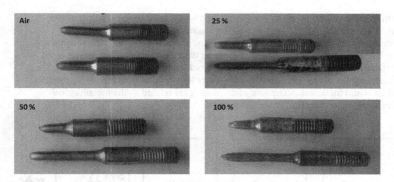

Figure 5. Images of the SSR specimens after been fractured exposed to soils with different moisture content.

Fig. 6 shows a SEM image from longitudinal sections of samples tested in the saturated calcareous soil, showing the presence of some typical secondary cracks close to the surface failure (primary crack). Fig. 6a shows the primary crack of the fracture surface. Fig. 6b shows the presence of some secondary cracks in the gage section close to the primary crack. The appearance of the fracture is typical of brittle surface. These crack seems to have the origin from a pit on the surface. However, not all pits were associated with a micro cracks. For the samples tested in the calcareous soil with 25 and 50% moisture content secondary cracks were not observed.

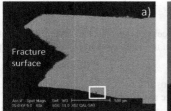 

Figure 6. SEM images of the longitudinal sections of samples tested in saturated calcareous soil showing a) the primary crack, b) secondary crack close to surface failure.

SCC process seems to be initiated from the bottom of a pit by a combination of dissolution and micro-stress on the pit (mechanical process). After the pit is generated by anodic dissolution, the pit continues growing and the geometry changes to a depth pit changing the aspect ratio of depth-wide, evolving in a depth and narrow pit bottom, which in turn intensifies the stress on the bottom. Micro cracks are initiated at the weakest link sites in the microstructure (in many cases start in the grain boundaries). The micro crack is formed due to a sharp crack tip, which then propagates in the bottom of the pit, thereby changing the pit into a sharp crack. This process is illustrated in Fig. 7.

According to Fang *et* al. [34], the pit geometry affects development of cracks. Smaller pits were less likely to crack. If the depth to width aspect ratio was <0.5, then the pits were less likely to crack. In addition, pit to crack transition was related to the severity of the loading conditions, corrosivity of the solution and the microstructure of the steel.

The morphology of some deep pits was semi circular at the bottom, which could not generate high stress concentration compared to those with sharper tips. Determination of critical pit size to transition from pitting to fatigue crack nucleation was developed by Shi *et al.* [35]. In addition, it is very important to considerer the metallurgical factors like phases presents, defects and inclusions. The size of the damage is related to the metallurgical properties of steel and to the corrosive environment.

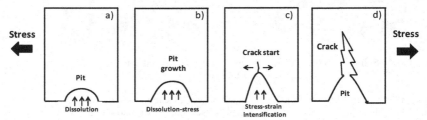

Figure 7. Schematic representation of the pit to crack transition.

The initiation of cracks is attributed to stress intensification on pits, which results in rupture of the passive film, thus exposing bare metal to the corrosive action of the environment. The stress level must be above a certain threshold level to produce a sufficient strain rate to exceed the passivation rate.

CONCLUSIONS

The investigation carried out on API X52 pipeline steel through SSRT in order to assess the susceptibility to SCC in calcareous soil with different moisture content permit to drawn the following conclusions:

- Physicochemical characterization of calcareous soil reveals a high carbonate content with a pH around 8, containing calcite, coesite and magnetite mainly.
- The results of RAR, TFR and PER, indicate that X52 pipeline steel was susceptible to SCC only in the calcareous soil saturated. SEM observations corroborate this asseveration, showing evidence of secondary cracks in the gage section.
- It was observed that some cracks were initiated from the bottom of pits. SCC can be initiated from the bottom of a pit by either a dissolution process or a mechanical process. Micro cracks are initiated at the weakest link sites in the microstructure.
- It can be attribute that a better dissolution of ions contained in the soil saturated enhance the corrosion activity giving origin to pits, and some of these pits with time evolved like cracks.

REFERENCES

1. P. Cazenave, S. Tandon, M. Gao, R. Krishnamurthy, R. Peverelli, C. Moreno, E. Díaz, Assessment and management of SCC in a liquid pipeline: case study, Proceedings of the 8th International Pipeline Conference (IPC-2010) Sept. 27-Oct. 1, 2010, Calgary, Alberta, Canada, Paper No. 31140.
2. B.N. Leis and R.J. Eiber, *Stress-Corrosion Cracking On Gas-Transmission Pipelines: History, Causes, and Mitigation*, Proceedings of First International Business Conference on Onshore Pipelines, Berlin, (1997.).
3. National Energy Board (NEB), *Report of the Inquiry-Stress Corrosion Cracking on Canadian Oil and Gas Pipelines*, (1996).
4. M.A. Arafin, J.A. Szpunar, *Materials Science and Engineering A*, **528**, 4927 (2011).
5. T. Omura, H. Amaya, H. Asahi, M. Sawamura, M. Kimura, *Corrosion,* Paper No. 09092 (2009)
6. M. Sawamura, H. Asahi, T. Omura, H. Kishikawa, N. Ishikawa, M. Kimura, *Corrosion,* Paper No.11286, (2011).
7. H. Asahi, T. Kushida, M. Kimura, H. Fukai, S. Okano, *Corrosion*, **55**, 644 (1999).
8. J.G. González Rodríguez, M. Casales, V.M. Salinas Bravo, J.L. Albarrán, L. Martínez, *Corrosion*, **58**, 584 (2002).
9. J.A. Beavers, B.A. Harle, *Journal of Offshore Mechanics and Arctic Engineering*, ASME, **123**, 147 (2001).
10. R.N. Parkins, *Stress Corrosion Cracking*, Uhlig's Corrosion Handbook, second edition, Edited by R. Winston Revie pp.191 (2000).
11. R.N. Parkins, W.K. Blanchard, B.S. Delanty, *Corrosion*, **50**, 394 (1994).
12. M. Elboujdaini, Y.Z. Wang, R.W. Revie, Initiation of stress corrosion cracking on X-65 linepipe steels in near-neutral pH environment, International Pipeline Conference (IPC) ASME, 967, (2000).
13. E. Shigeru, N. Moriyasu, K. Yasuo, U. Kazuyoshi, *ISIJ International*, **34**, 217 (1994).
14. B.W. Pan, X. Peng, W.Y. Chu, Y.J. Su, L.J. Qiao, *Materials Science and Engineering A*, **434**, 76 (2006).
15. J. Bulger, J. Luo, *Effect of microstructure on near-neutral pH SCC*, International Pipeline Conference (IPC) ASME, 947, (2000).
16. B.S. Delanty, J. O'Beirne, *Oil & Gas Journal*, **15**, 39 (1992).
17. Z. Szklarska, Z. Xia, R.B. Rebak, *Corrosion*, **50**, 334 (1994).
18. NACE TM-0198 Slow Strain Rate Test Method for Screening Corrosion-Resistant Alloys (CRAs) for Stress Corrosion Cracking in Sour Oilfield Service, (2004).
19. G. M. Ugiansky, J.H. Payer (Eds.), Stress Corrosion Cracking—The Slow Strain Rate Technique, American Society for Testing and Materials, Philadelphia, ASTM STP 665, (1979).
20. R. D. Kane, C.J.B.M. Joia, A.L.L.T. Small, J.A.C. Ponciano, *Materials Performance*, **36**, 71 (1997).
21. ASTM G-129, *Slow strain rate testing to evaluate the susceptibility of metallic materials to environmentally assisted cracking*, (2013).
22. A. Contreras, A. Albiter, M. Salazar, R. Pérez, *Materials Science and Engineering A*, **407**, 45 (2005).
23. A. Contreras, S.L. Hernández, R. Galván, *Mater. Res. Symp. Proc.*, **1275**, 43 (2011).
24. Z. Velázquez, E. Guzmán, M.A. Espinosa, A. Contreras, *Mater. Res. Soc. Symp. Proc.*, **1242**, 69 (2010).
25. A. Contreras, M. A. Espinosa Medina, R. Galván Martínez, *Mater. Res. Soc. Symp. Proc.*, **1275**, 53 (2011).

26. B. Y. Fang, A. Atrens, J. Q. Wang, E.H. Han, Z.Y. Zhu, W. Ke, *J. of Materials Science*, **38,** 127 (2003).
27. R.N. Parkins, R.R. Fessler, *Materials in Engineering Applications*, **1,** 80 (1978).
28. B. Gu, W.Z. Yu, J.L. Luo, X. Mao, *Corrosion,* **55,** 312 (1999).
29. A. Contreras, S.L. Hernández, R. Orozco Cruz, R. Galván Martínez, *Materials & Design*, **35,** 281 (2012).
30. A. Contreras, E. Sosa, M.A. Espinosa, *Mater. Res. Soc. Symp. Proc.,* **1242,** 43 (2010).
31. T. R. Jack, B. Erno, K. Krist, R.R. Fessler, NACE International Corrosion, Paper No. 00362 (2000).
32. A. Benmoussat, M. Hadjel, *Journal of Corrosion Science and Engineering*, **7,** 1 (2005).
33. Z.Y. Liu, X.G. Li, C.W. Du, L. Lu, Y.R. Zhang, Y.F. Cheng, *Corrosion Science*, **51,** 895 (2009).
34. B.Y. Fang, R.L. Eadie, W.X. Chen, M. Elboujdaini, *Corrosion Engineering Science and Technology*, **45,** 302 (2010).
35. P. Shi, S. Mahadevan, *Engineering Fracture Mechanics*, **68,** 1493 (2001).

Characterization of Materials for Industrial Applications

Mater. Res. Soc. Symp. Proc. Vol. 1766 © 2015 Materials Research Society
DOI: 10.1557/opl.2015.417

Characterization of Mg AZ31 Alloy ECASD Processed Using Dynamical Mechanical Analysis (DMA)

Daniel Peláez[1], Adriana Restrepo-Osorio[2], Emigdio Mendoza[2], Cesar Isaza[3], Patricia Fernandez-Morales[4]

E-mail: patricia.fernandez@upb.edu.co

[1] Facultad de Ingeniería Mecánica, Universidad de Colombia.
[2] Facultad de Ingeniería Textil, Universidad de Colombia.
[3] Universidad Nacional de Colombia-Sede Medellín.
[4] Facultad de Ingeniería Industrial, Universidad Pontificia Bolivariana, Colombia.

ABSTRACT

In the present work was used Dynamical Mechanical Analysis (DMA) to study the magnesium alloy Mg AZ31-B, in plate form with a thickness of 2.5 mm. The plates were processed using Equal Channel Angular Sheet Drawing (ECASD), which is a severe plastic deformation technique, which allows imposing strain without dimensional changes to a metal plate, at room temperature with an angle of 135°. The obtained results show dependence between the storage modulus (M'), temperature and frequency used on the tests. The greater M' values were obtained at the lower temperatures and at the higher frequency used. However, at lower frequencies M' response is not affected by the used frequencies. At the higher temperatures there is an M' reduction, which promotes the material deformation.

Keywords: Mg, alloy, phase transformation, absorption, steel.

INTRODUCTION

Dynamical Mechanical Analysis (DMA) can be described as applying an oscillating force to a sample in a temperature range, and then analyzing the response to that imposed force. The advantage of this technique is that allows studying the material's mechanical properties as a temperature and a frequency function. This technique has been widely used for polymers characterization due to its sensitivity to weak transitions [1, 2]. Recently, dynamical mechanical analysis has been used in metals for analyzing the damping capacity [3-5], and in alloys to study the phase's transformation [6-9].

The damping capacity is defined by the tan δ, which quantifies the way in which a material absorbs and disperses energy. It expresses the out-of-phase time relationship between an impact force and the resultant force that is transmitted to the supporting body. The tan δ is ultimately an indication of the effectiveness of a material's damping capabilities. The higher the tan δ, the greater the damping coefficient, the more efficient the material will be in effectively accomplishing energy absorption and dispersal.

Equal Channel Angular Sheet Drawing (ECASD) is a severe plastic deformation technique, which allows imposing strain without dimensional changes to a metal plate. ECASD has been recognized as a simple and effective technique for improving materials mechanical properties; among the most studied alloys using SPD techniques are magnesium alloys [10-12], and in a smaller proportion other alloys such as ferrous alloys [13, 14], aluminum alloys [15, 16], copper alloys [17, 18] and titanium alloys [19, 20].

Nowadays are needed materials, which can simultaneously exhibit a high, damping capacity and high specific mechanical properties. High damping capacity materials allow to suppressed passively mechanical vibrations and wave propagation.

Nevertheless, having both properties in metals is almost incompatible due to the microscopic mechanisms involved in damping and strengthening. For the above reason, it is interesting to develop materials that simultaneously exhibit good mechanical properties and high damping [21-23]. Mg alloys exhibit a higher specific strength compared to other alloys such as aluminum alloys or some types of steels [24, 25]. For this reason, Mg alloys have been widely characterized under several loading conditions [26-28].

However, dynamic properties characterization of Mg alloys at high temperatures has received less attention; nevertheless, those are critical in the manufacturing process.

Hence, the aim of this study is to evaluate the changes on storage modulus (M'), which is related to the mechanical properties of the material, and tan δ, also known as damping capacity, as a function of the temperature and frequency in an AZ31 magnesium alloy when is processed by ECASD with an angle of 135°. This process has been carried out at room temperature and up to six passes of shear using route A [29].

EXPERIMENTAL

Magnesium alloy AZ31-B sheets in annealed condition, with a nominal thickness of 2.4 mm, were used. Samples were processed by ECASD (Fig. 1) at room temperature (RT). The angle formed by the intersecting channels is 135°. A universal testing machine INSTRON 5582 was used for passing the samples through the die at a processing speed of 20mm/min, and to six shear passes using route A, in which the specimen remains the same orientation about its longitudinal axis between successive passes.

The dynamic mechanical analysis was performed using DMA/SDTA 861 Mettler Toledo equipment. The equipment has a temperature range of -150 to 500 °C and a load cell of 18N. The tests were conducted in the single cantilever mode. Rectangular specimens were used with the following dimensions: length 28.46 mm, width 3.88 mm, thickness 2.4 mm. The temperature scan was performed from RT to 400 °C, at a heating rate of 3 °C/min and using the following frequencies of 0.1, 1, 10, and 100 Hz. Load and displacement limits were 10N and 800 μm, respectively. The results are processed in the STARe software.

Figure. 1. ECASD setup.

RESULTS AND DISCUSSION

In Figure 2 and Figure 3 are presented the results for the DMA test. In the top of the figure is showed the storage modulus (M') and on the bottom tan δ. Each of the curves is presented as a function of the series of frequencies used and for a temperature sweep from RT to 400 °C. Figure 2 presents the results for the as-received material, it means without any processing. It is observed that the storage modulus curves have a similar behavior for every tested frequency, being subtly different for the 100 Hz frequency, for the latter the maximum storage modulus is reached. However, for low frequencies, 0.1, 1 and 10 Hz, is observed an almost perfect curves overlap up to 125°C, above this temperature there is a difference between 0.1 and 10 Hz frequency curves of almost 0.5 GPa. At 320 °C the curves for all the four tested frequencies overlap again up to 400 °C, temperature in which is obtained the lower M' value. For all the tested frequencies is observed that at 140 °C is given a change in the curve slope, and there is when begins to be more noticeable the increase in the damping capacity, tan δ. This can be due to being close to the re-crystallization temperature of the material, and hence the exhibited stiffness by the material decreases. Above 320 °C the curves at the different frequencies begin having the same storage modulus value, showing that the material at this temperature is already in a like fluid state, and therefore its behavior is no longer dependent of tested frequency.

In the bottom of Figure 2 is presented the results for tan δ for each of the tested frequencies. Similarly to the storage modulus curves, it is seen that the curves tendency for all the frequencies is similar. However, for the 0.1 Hz frequency above 200 °C begins to increase its value, no longer overlapping with the 1, 10 and 100 Hz frequencies. At 120 °C is observed for all frequencies and increase in the tan δ value. Nevertheless, the 100 Hz frequency from 300 to 350 °C shows a diminution in the tan δ value; after 350 °C, its value increases again but being less than for the other frequencies. Around 200 °C there is a slope change for all tested frequencies, this slope change is due to the phase change from $(Mg) + \beta \rightarrow (Mg)$, which n the phase diagram for the AZ31 is reported to occur around 210°C.

The increase in the values of tan δ could be related to an increase in the material damping capacity [30].

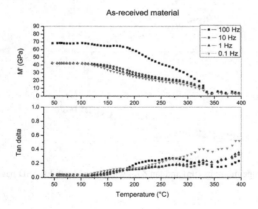

Figure 2. Variation of storage modulus (up) and tan δ (down) for the as-received material.

The results for the storage modulus and for tan δ for the ECASD processed sample are shown in Fig. 3. The storage modulus curve shows a similar behavior than the presented above for the as-received state material. The curves for the 0.1, 1 and 10 Hz frequencies are overlapped up to 125 °C, showing that up to that temperature the change in the storage modulus is not related to the used frequency, being only affected by the temperature. For the 100 Hz frequency is seen a direct frequency influence on the storage modulus, getting the highest values at this frequency. Above 360 °C is observed a similar behavior for all frequencies, having all the curves the same M' values, because the like fluid behavior becomes dominant in the material response. For the ECASD processed sample is observed that the temperature at which the storage modulus is independent of the tested frequencies is higher than this for the material in the as-received state. This may be because the ECASD processing modifies the crystal structure of the material altering the crystallographic planes and thus altering the mechanical properties exhibited by the material [16, 15]. The value of the storage modulus for the ECASD processed material is greater, being almost twice the value of the material in as-received state, this is because the ECASD process favors the grain refinement and hence the increase in the mechanical properties [11, 20].

Results for tan δ for the ECASD processed sample are presented in the bottom of Fig. 3. In this graph the differences is more noticeable when compared to the as-received material. For the 100 Hz frequency the curve exhibited a higher value when compared to the other frequencies. At 275 °C, the 0.1 Hz frequency overcomes the 100 Hz frequency, and above 300 °C the 100 Hz frequency, tan δ values are less than those for the other frequencies. For the ECASD processed sample the tan δ values are higher than those for the as-received sample, which can be related to an increase in the plasticity exhibited by the material. It is related to finer grain sizes and an accommodation thereof [30].For the ECASD samples is not visible the slope change due to the phase transformation around 200°C, as reported by Mingler et al [33] this is due because of the Equal Channel Angular process.

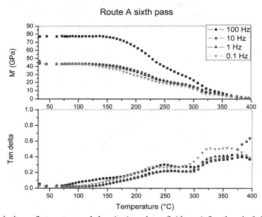

Figure 3. Variation of storage modulus (up) and tan δ (down) for the sixth ECASD pass.

For both cases the higher damping capacities were achieved for the 0.1 Hz frequency because the stress applied in the measurement is very small; then, the interfacial slip can only be called as micro-slide. It is well known that the micro-slide can accomplish more exhaustively at a low frequency rather than at a high frequency under the same stress level because frequency is the reciprocal of the stress's cyclic time. This resulted in another phenomenon that damping increases with decreasing the frequency [4]. Moreover, Mg alloys has a remarkable dislocation damping effect. The dislocation damping effect is caused by the movement of dislocations, which are weakly pinned on the basal plane [3], which is seen on the damping frequency capacity for the tested frequencies.

The damping capacity for the as-received material at 1 Hz is higher than the value reported for Hazeli *et al.* [5]; nonetheless, they used extruded bars while this study used annealed rolled plates, showing that the processing technique could also have an influence in the initial damping capacity.

CONCLUSIONS

Dynamical mechanical analysis has been carried out on ECASD processed magnesium AZ31 alloy at six passes at room temperature for route A. The ECASD process at the six pass has an evident effect in increasing the storage modulus of the material at lower temperatures, being almost the double for the initial temperature, when is compared to the as-received material. The increase in the storage modulus is related to the decrease in the grain size because according to the Hall-Petch relation, the decrease in the grain size is related with the increase in the yield strength.

Nevertheless, tan δ has no changes in its magnitude for both cases; however, the better damping capacities are achieved for lower frequencies. Above 340°C the value for the storage modulus is equal in magnitude for the two tested conditions, showing that above that temperature de storage modulus is no longer frequency dependent.

This could be related to the material is in a viscous state because of the temperature values, which means that its behavior is more like a fluid than like a solid. The as-received material has no major changes as a function of the frequency but the ECASD sample has a very different behavior for the 100 Hz series; however, at the lower frequencies the material behavior is similar. The results show that the ECASD process is a feasible method for processing Mg alloys and obtaining high specific strength materials with an enhanced damping capacity over the tested temperature range.

ACKNOWLEDGMENTS

To Universidad Pontificia Bolivariana, Universidad Nacional de Colombia Sede Medellin and to Colciencias who financially supported this work with project code 1210-569-34713.

REFERENCES

1. K. Menard, Dynamic Mech. Analysis. A practical Introduction, CRC Press, Boca Raton, (1999).
2. L. Nielsen L, R. Landel, Mech. Properties of Polymers and Composites, CRC Press,Boca Raton, (1994) .
3. H. Lu, X. Wang, T. Zhang, Z. Cheng, Q. Fang, *Mater,* **2,** 958 (2009).

4. J. Gu, X. Zhang, M. Gu, *Mater. Trans.*, **45**, 1743 (2004).

5. K. Hazeli, A. Sadeghi, M. Pekguleryuz, A. Kontsos, *Mater. Sci. Eng. A*, **589**, 275 (2014).

6. P. Buechner, D. Stone, R. Lakes, *Scr.Mater,* **41**, 561 (1999).

7. G. Süleyman, *Turkish J. of Eng.& Environ.Sci.*, **26**, 353 (2002).

8. J. Van Humbeeck, *Journal of Alloy. and Compd*, **355**, 58 (2003).

9. Q. Wang, H. Fusheng, J. Wu, H. Gangling, *Mater. Lett*, **61**, 2598 (2007).

10. F. Zhao, Y. Li, T. Suo, W. Huang, J. Liu, *Trans of Nonferr Met. Soc. of China*, **20**, 1316 (2010).

11. Y. Kim, D. Yang, *Int. Journal of Mech. Sci.*, **27**, 487 (1985).

12. Y. Cheng, Z. Chen, W. Xia, *Mater. Charact*, **58**, 617 (2007).

13. L. Ma, X. Wu, K. Xia, *Mater. Forum*, **32**, 35 (2008).

14. M. Eddahbi, M. Monge, T. Leguey, P. Fernández, R. Pareja, *Mater. Sci. and Eng.*, **528**, 5927 (2011).

15. Y. Chen, Y. Chai, S. Gireesh, J. Hjelen, *Mater. Sci. and Eng. A*, **545**, 139 (2012).

16. U. Chakkingal, A. Suriadi, P. Thomson, *Mater. Sci. and Eng. A*, **266**, 241 (1999).

17. D. Chen, J. Wang, G. Tzou, *Mater. Sci. Forum*, **594**, 90 (2008).

18. N. Stepanov, A. Kuznetsov, G. Salishchev, G. Raab, R. Valiev, *Mater. Sci. and Eng. A,* **554**, 105 (2012).

19. X. Zhang, L. Cao, Y. Zhao, Y. Chen, X. Tian, J. Deng, *Mater. Sci. and Eng. A,* **560**, 700 (2013).

20. X. Zhao, X. Yang, X. Liu, X. Wang, T. Langdon, *Mater. Sci. and Eng. A,* **527**, 6335 (2010).

21. J. Zhang, R. Perez, E. Lavernia, *Journal Mater. Sci.,* **28**, 2395 (1993).

22. N. Srikanth, X. Zhong, M. Gupta, *Mater. Lett.,* **59**, 3851 (2005).

23. R. Schaller, *J. Alloys Compd.*, **355**, 131 (2003).

24. R. Gehrmann, M.M. Frommert, G. Gottstein, *Mater. Sci. and Eng. A*, **395**, 338 (2005).

25. K. Hazeli, A. Sadeghi, M. Pekguleryuz, A. Kontsos, *Mater. Sci. and Eng. A,* **578**, 389 (2013).

26. W. Miller, L. Zhuang, J. Bottema, A. Wittebrood, P. DeSmet, A. Haszler, *Mater. Sci. and Eng. A*, **280**, 37 (2000).

27. M. Tmuka, Y. Watanabe, K. Higashi, *Scripta Mater,* **45**, 89 (2001).

28. M. Barnett, Z. Keshavarz, A. Beer, D. Atwell, *Acta Mater*, **52**, 5093 (2004).

29. W. Wu, S. Lee, A. Paradowska, Y. Gao, P. Liaw, *Mater. Sci. and Eng. A,* **556**, 278 (2012).

30. C. Boehlert, Z. Chen, I. Gutiérrez-Urrutia, J. Llorca, M. Pérez-Prado, *Acta Mater,* **60**, 1889 (2012).

31. V. Segal, *Mater. Sci. and Eng. A*, **197**, 157 (1995).

32. E.E Martínez-Flores, J. Negrete, G. Torres-Villaseñor, *Rev. Matéria*, **13**, 365 (2008).

33 B. Mingler, O.B. Kulyasova, R.K. Islamgaliev, *Nanostructured materials-processing, structures, properties and applications*, **42**, 1477 (2007).

Mater. Res. Soc. Symp. Proc. Vol. 1766 © 2015 Materials Research Society
DOI: 10.1557/opl.2015.418

Chemical Characterization of a Mineral Deposit of Economic Interest

E. Cerecedo, V. Rodríguez, P.D. Andrade, E. Salinas, J. Hernández, A. Arenas

Área Académica de Ciencias de la Tierra y Materiales. Universidad Autónoma del Estado de
Hidalgo. Carr. Pachuca – Tulancingo km. 4.5 Mineral de la Reforma Hidalgo, México. C.P.
45184 Tel. (01 771) 71 720 00 ext. 2279
E-mail mardenjazz@yahoo.com.mx

ABSTRACT

Chemical and structural characterization of four representative samples of an ore deposit located in the eastern of Hidalgo State was carried out. According with the results, it could
be appreciate some areas showing silicified zones with abundant amounts of disseminated py-
rites that are part of a rock unit from early Jurassic consisting in inter - bedded black shales
and sandstones. Thus, the contents of base metal were greater than 30 ppm Zn and 9 ppm Cu.
Chemical analysis of rock gave the following results; 82 ppm of Ba, 1.64 % Wt. Fe, 0.08 %
Wt. Ti, 40.8 % Wt. Si, 20 ppm of Ce, 2.2 ppm Co, 30 ppm Cr, 2.7 ppm Cs, 0.9 ppm Er, 2.5
ppm Ga, 1.6 ppm Gd , 1.5 ppm Ge, 9 ppm La, 71 ppm Li, 104 ppm Mn, 10 ppm Nd , 17 ppm
Rb, 2 ppm Se, 9 ppm Sr, 10 ppm Ta, 6 ppm Te, 28 ppm V, 9 ppm Y, and 0.7 ppm Yb,
among others. Finally, the values found for precious metals, were; Au < 0.02 ppm, Pd <0.05
ppm, Pt <0.05 ppm. It was inferred that the low content of base metals in outcrop studied, are
due to the alteration of the black shales. According to these results, we can consider a stratiform – type mineralization of Pb-Zn which could be prospective for SEDEX – Type deposit.
By means of XRD, it was possible to identify; pyrite, chalcopyrite, pyrrhotite, and minor
amounts of sphalerite and Co -Ni arsenide.

Keywords: Structural, metal, X-ray Diffraction (XRD), core, Scanning Electron Microscopy
(SEM).

INTRODUCTION

The sedimentary – stratiform deposits of Pb – Zn of sedimentary exhalative type kind
(SEDEX), contain more than 50 % of Pb and Zn world reserves [1], and they occur in sedimentary archaic sequences of heterochronic ages.

On the other hand, one of the principal questions during prospecting of SEDEX – type
deposits, is what kind of predominant environment was during their formation; which could
be oxidizing or reducing and this is determinant to describe the affinity of deposit according
initial estimated tonnage and metal content of deposits.

Although some studies have revealed [2] that when shales sequences and siliclasts intervals exist, in a general manner the sequence could be classified as reducing environment
and their contents can be considered according to the correlation between works related to
this kind of deposits showing high metal base content and medium tonnage that can be classified as SEDEX deposit of Shelwyn sub – type [1]. Meanwhile, giant deposits with oxidizing
lithology are Mcarthur type.

EXPERIMENTAL

Drilling cores was obtained in the ore deposit to evaluate the quality of material. Samples so obtained were completely characterized using different analytical methods to establish a preliminary systematic analysis of this deposit; so efforts were concentrated on the first four meters depth of principal mineralized zone, observing by mesoscopic analysis in the deposit, an outcrop of fine disseminated pyrite and high prevalence of silicification.

On the other hand, Scanning Electron Microscopy was used to study texture and semi quantitative distribution of elements on drilling samples. For this, samples used to determinate semi quantitative composition by Energy Dispersive Spectrometry (EDS) were grinded to + 200 mess (+ 35 μm average size) and mounted in an aluminum sheet adhesive to the analysis. Then core samples were cut and covered with a gold film to analyze texture in a JEOL JSM-6300 microscope, equipped with an EDS detector.

Sample preparation to X – Ray Diffraction analysis (XRD) was carried out by grinding to a -10 μm average size and compacting sample, so obtained powder was put in an aluminum sample holder of 1 cm diameter, using an Al piston of 2.54 cm length. Finally, sample holder was put in an INEL diffractometer, Equinox 2000 model provided with a curved detector operable with an ethanol – argon gas mixture of high pureness (99.9 %). XRD results were treated with diffraction software Match 3.0 to index mineral species present in samples. Finally, to determinate chemical composition of samples, Spectrometry of Inductive Couple Plasma (ICP) and Mass Spectrometry of ICP were used by Na_2O_2 fusion for rock analysis, and fire assay – ICPOES for platinum, palladium, and gold. The latter analyzes were carried out at ActLab LTD in Canada.

RESULTS AND DISCUSSION

In the zone where core samples were extracted, it was possible to observe a continuous stratigraphic section that allows to obtain a preliminary relationship of geology, stratigraphy and tectonic of place to suit the studied mineralization in a relative age of lower Jurassic [3].

Stratigraphic sequence where from drilling cores were extracted, corresponds to lower Jurassic and stratigraphic column composition obtained by SEM – EDS shows the existence of two exhalative pulses that can be inferred by the presence of high temperature minerals observed by XRD and with major concentration of exhalative minerals type $(Mg,Fe)Fe_2O_4$, in contrast with minerals of medium to low temperature which are indicative of the end of pulse, in this case like albite $NaAlSi_3O_8$ and other tecto - silicates.

It is also inferred that such pulses are referred to 1 to 3 meters depth with intervals from pulse to pulse of about 10 cm thick.

Meanwhile, there is a substantial difference in calcium content of pulse of 1 and 3 meters; where at 1 meter higher counts to 1125 are obtained of calcium phase and at 3 meters, these counts decrease to one tenth.

In the same way, it is evident that mineralization is more amorphous at 3 meters depth where this presents an anomaly of hematite of botryoidal type which is typical of exhalative emanations to which we can attribute the change in crystalline shape from cubic and tetragonal to an amorphous one. Also, it is important to note the immiscibility of silicate phase and carbonate phase which is shown through column lithological.

Likewise, by XRD analysis was identified the relative abundance of at least 4 predominant minerals in the studied deposit. These are; Quartz (PDF: 00-046-1045), Albite (00-003-0508), Calcium carbonate (01-072-1616), Aluminum oxide (00-050-1496) and pyrite (01-

076-0964). While other minerals such as; Chalcopyrite (00-037-0471), Pyrrhotite (00-002-1241) and sphalerite (00-036-1450), in addition to small amounts of orcelite of Co – Ni arsenide [orcelite (Ni, Fe, Cu)$_{4.2}$ (As,S)$_2$] as shown in figure 1.

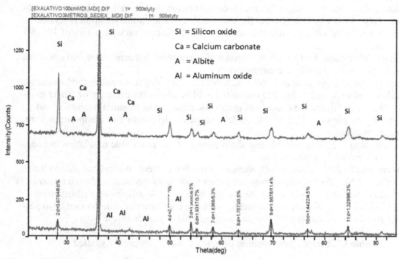

Figure 1. XRD Spectrum showing main phases of silicon oxide and calcium carbonate

In the studied deposit, predominant silicates are quartz and albite and also, small quantities of chlorite was observe. Figure 2 shows an albite crystal that corresponds to mineralization zone at medium temperature.

 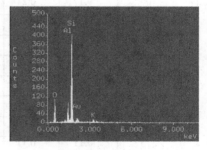

Figure 2. Particular image by SEM and EDS spectra of sample, showing an Albite Crystal

The occurrence of albite crystals exhalites is ambiguous, and is principally distributed between exhalative pulses. In the figure, it also can be observed at the centre, a prism 120 μm length and 60 μm thick. EDS analysis, suggests a tectosilicates minerals intergrowth in drusen brecciate areas associated and possibly, the presence of sodium chloride detected in some samples analyzed by EDS comes from connate water.

Noteworthy that by means of elements concentration profile obtained in the mineralized zones (Table 1), it was inferred that the coincidence of such mineralization is related with effervescence zones in a particular time like is shown in column with pulses of 10 – 50 cm in the rock.

On the other hand, Fe_2O_3 content is essentially constant, but arriving at 300 cm depth increases sharply and gradually decreasing depth.

MgO content is pulsatile at 1 and 3 meters essentially and it is evident the association of magnesium and iron, and maybe the anomaly could be due to the presence of ferrite magnesium [$(Mg,Fe)Fe_2O_4$], but we should not ignore the presence of titanium magnetite and perhaps, some pure phase of Ti, possibly rutile because dendrite nature of components from lithological formation.

At the same way, it is possible that the presence of calcium oxide in relation to Si appears to be an immiscibility mechanism.

According to XRD studies, both alumina and silica contents perhaps are due to the presence of albite $NaAlSi_3O_8$, where these are present like tecto – silicate and such presence is quite illustrative because its origin is through sequences of felsic affinity by hydrothermal processes like chilling exolution, that coincides with the cool interpulsatile portion, also perhaps deficiency of S in column could be due to a bacterial – genetic effect.

Table I. Chemical composition of mineral species in profile studied

Depth (cm)	Fe_2O_3 (ppm)	MgO (ppm)	Al_2O_3 (ppm)	SiO_2 (ppm)	CaO (ppm)
- 50	0.14	0.12	1.67	97.31	0.19
- 75	0.00	0.00	6.18	97.45	0.07
- 100	0.17	0.21	1.55	91.84	0.00
- 125	0.00	0.00	2.82	97.89	0.00
- 250	0.00	0.18	5.82	96.64	0.00
- 275	0.18	0.24	4.23	92.55	0.05
- 300	2.64	0.55	0.99	90.80	0.00
- 325	0.29	0.00	1.41	97.37	0.00
- 350	0.02	0.00	4.70	98.04	0.03
- 375	0.11	0.11	3.33	93.96	0.05
- 400	0.00	0.00	0.44	96.02	0.02
- 425	0.00	0.00	2.78	98.99	0.00
- 450	0.00	0.08	0.03	96.61	0.03

Geochemistry of trace elements

ICP analyzes of collected samples, were treated to compare the abundance of elements between mineralized and underlain areas without mineralization, and a chondrite sample [4], (Figure 3).

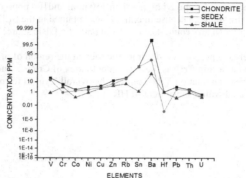

Figure 3. Multi elements diagram used to compare chondrite value (UCC Mclean)

In the geochemical analysis of trace elements such as Cr and Th, due their low mobility, these are considered suitable for determining provenance and tectonic setting [5]. As seen from the diagram, that is an enrichment of geochemical elements large ion such as Cr =30 and Th = 1.8; all higher than the average for simple of SEDEX deposit in comparison with that values from continental crust and certainly, Huayacocotla shale barren, so it follows that the rock that caused the lithological formation is close to the source area.

On the other hand, the contents of transition trace elements V = 28, Cr = 30, Co = 2.2, Cu = 7 and Ni = 10, are low due to affinity of SEDEX deposit is felsic or acid.

Indeed, while the content of Cr and Ni is low which is indicative for a felsic source, but it presents an anomaly with high values of V = 28, Co = 2.2 and Cu = 7. Such anomaly for V could be associated to contents of metals from platinum group.

Also this SEDEX deposit has higher concentration of light rare earths (LREE) than chondrite, because values for heavy rare earths (HREE) are low for Y = 8.8 and Ga = 2.5. For this reason, it could be inferred the presence of associated minerals of REE of the monacite - type [(Ce,La,Nd, Th, (PO)$_4$] and bastnaesite minerals [6] that produce almost the 95 % of REE actually.

It is important to note that negative anomaly of HREE for Eu and Hf with respect to chondrite, confirms the impoverishment of HREE and reaffirms the felsic affinity of SEDEX deposit and possibly its origin from a Gneissic protolith because mineralization has higher concentrations of LREE.

In the diagram Cr/V vs. Y/Ni of figure 4 can be observed a wide difference between different samples of shales from Huayacocotla, the studied SEDEX deposit and the chondrite.

It can be identified two regions divided by a diagonal line that correspond to the felsic rocks (granite) y mafic (basalt) respectively and values of Cr/V are low in SEDEX deposit and then are situated in the felsic area.

On the other hand, relationship of Y/Ni against Cr/V shows a felsic affinity for analyzed samples.

According [7] contents of Cr and Ni higher than 150 ppm and100 ppm, respectively, are indicative of a ultramafic source, because maxim values obtained in the SEDEX deposit are of 30 ppm Cr and 10 ppm Ni and could be considered part of a felsic source for SEDEX deposit.

To establish the relative proportion of ultramafic rock in the region of origin, it is important to point that the relationship Cr/V is the enrichment index of Cr over trace ferro magnesian elements. Meanwhile, the relation Y/Ni controls the overall level of ferro magnesian trace elements (Ni) compared with a substituent of HREE.

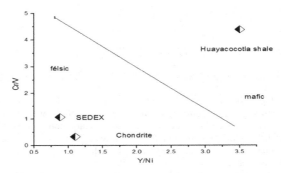

Figure 4. Diagram Cr/V *vs.* Y/Ni for identification of felsic and mafic rocks

Mineralized body shows clearly a continuous mineralogical zoning, which is reflected in analyzes of Au, Pt, V, Cr, Cu, Ag, Zn, Pb and Ba. According this, gold is preferentially enriched when a lower silidification is present as the same way like platinum; also, a homogeneous distribution of gold is perceived on the pulse, but with lower concentration on the more superficial portion. Meanwhile, vanadium and chromium are related to the magnesium content.

Copper and silver have cross predominance enrichment; meanwhile, the zinc is enriched near the surface. Likewise, lead and barium have similar behavior zonal.

With profile elements, it follows that the mineral composition is heterogeneous, so from the elements only it is possible to infer the possible occurrence of minerals, so the barite would expect to be restricted in the upper portion of mineral body joint with galena and sulfosalts. And expect sphalerite was located preferentially in the in the upper portion, but most probably is located throughout the body like pyrite. Meanwhile, expect chalcopyrite is restricted to the lower body.

The silicification would also be restricted to the lower part of the body and interpulsatile. And barite, galena quartz and chalcopyrite are preferentially in the midsection.

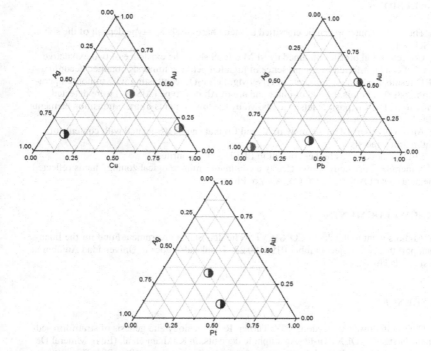

Figure 5. Ternary diagrams of composition for SEDEX deposit at depth.

Additional information of depth distribution for certain elements was done by SEM – EDS analysis, plus a comparison using ternary diagrams. There is a strong contrast at depth in the composition of diagrams of Au-Ag-Pt-Cu-Pb, of figure 5, as to the contents of lead and copper, increasing the copper content of depth; but a more or less homogeneous behavior in gold and platinum content depth is observed. Perhaps, the contrasting change in the elements is due to a lenticular geometry of reservoir that would partly explain the depth distribution of the elements.

Although shales are indicative of an anoxic depositional environment, bacterial reduction is important because maybe they develop in these environments and can produce sulfides by sulfate reduction [8].

CONCLUSIONS

The studied outcrop can be classified as metal base SEDEX – type deposit of the sub type Selwyn.

Concentration profiles obtained by SEM – EDS show the existence of two exhalative pulses that can be inferred by the presence of high temperature minerals obtained according to XRD results. Exhalative minerals of type $(Mg, Fe)Fe_2O_4$, contrasting with minor temperature minerals that points the end of pulse and are of Albite – type $(AlSi_3O_8)$ were detected.

Affinity of SEDEX deposit is felsic, as shown by the low contents of trace transition elements such as V, Cr, Co, Cu, and Ni.

Also was observed an anomaly of V, and Co that can be associated with contents of platinum group metals.

SEDEX deposit has higher content in LREE that chondrite.

Mineralized ore body shows clearly a continuous mineralogical zoning, that is reflected by the analyzes of Au, Pt, V, Cr, Cu, Ag, Zn, Pb and Ba.

ACKNOWLEDGMENTS

Authors want to thanks to CONACyT – Hidalgo State Government Fund for the financial support through project number 0193180 22. Thanks also goes to Universidad Autónoma del Estado de Hidalgo.

REFERENCES

1. W.D. Goodfellow, J.W. Lydon, R.W. Turner, R.W, Geology and genesis of stratiform sediment-hosted (SEDEX) Zn-Pb-Ag sulphide deposits, *in* Kirkham et. al. (Eds). Mineral Deposit Modeling, Geological Association of Canada, Special Paper 40, p. 201-251, (1993).
2. D.R. Cooke, S.W. Bull, R.R. Large, P.J. McGoldrick, "The importance of Oxidized Brines for the Formation of Australian Proterozoic Stratiform Sediment-Hosted Pb-Zn (SEDEX) Deposits, Economic Geology. Vol 95, (2000).
3. S.E. Cerecedo, R.E. Salinas, "Guía rápida de Exploración Geológica Minera", ISBN 978-3-659-06942-0. Editorial Académica Española, Saarbrücken, Alemania, (2013).
4. N.M. Evensen, P.J. Hamilton, R.K. O'Nions, *Geochimica et cosmochimica acta*, **42, 8, 19,**1199 (1978).
5. S.M. Mclenan, S. Hemming, D.K. McDaniel, G.N. Hanson, *Geological Society of America, Special paper*, **284**, 21 (1993).
6. S. Jorge, D.J.C. Melgarejo, P. Alfonso, Asociaciones minerales en sedimentos exhalativos y sus derivados metamorficos. In Melgarejo, J.C.D. (Editor). "Atlas de Asociaciones Minerales en Lámina Delgada". Ediciones Universitat de Barcelona. Pp. 287 – 303. (1997).
7. J.I. Garver, P.R. Royce, T.A. Smick, *Journal sediment. Res.*, **66,** 100 (1996).
8. P.M. Okita, Manganese Carbonate Mineralization in the Molango District, Mexico, Economic Geology, **87**, 1345 (1992).

Mater. Res. Soc. Symp. Proc. Vol. 1766 © 2015 Materials Research Society
DOI: 10.1557/opl.2015.419

Effect of Composition on the Physical Properties of $(TlInSe_2)_{1-x}$ $(TlGaTe_2)_x$ Solid Solutions

M.M. Asadov[1*], S.N. Mustafaeva[2], D.B. Tagiev[1], A.N. Mammadov[1]

[1]Institute of Catalysis and Inorganic Chemistry, Azerbaijan National Academy of Sciences,
Pr. H. Javid 113, Baku, AZ1143 Azerbaijan
Email: mirasadov@gmail.com

[2]Institute of Physics, Azerbaijan National Academy of Sciences,
Pr. H. Javid 131, Baku, AZ1143 Azerbaijan

ABSTRACT

We carried out thermodynamic study of the $TlInSe_2$-$TlGaTe_2$ system based on the data of physicochemical analysis. Based on thermodynamic analysis and concentration dependence of physical properties, it was found that there is anion-cation substitution in $TlInSe_2$-$TlGaTe_2$ system. Continuous series of $(TlInSe_2)_{1-x}$ $(TlGaTe_2)_x$ solid solutions is forming throughout entire concentration range. We determined dielectric characteristics of samples, their frequency dispersion and nature of dielectric losses. The results demonstrate that the dielectric dispersion in the studied crystals $TlInSe_2$ and $(TlInSe_2)_{0.5}(TlGaTe_2)_{0.5}$ has a relaxation nature. Hyperbolic decline of loss tangent with increasing frequency from 50 kHz to 35 MHz indicates the loss of pass-through conduction in $(TlInSe_2)_{1-x}(TlGaTe_2)_x$ solid solutions.

INTRODUCTION

$TlInSe_2$ and $TlGaTe_2$ single crystals are typical representatives of chain-layered semiconductors and attract a lot of attention due to their interesting physical properties. These properties include strong anisotropy of the electric parameters related to special features in the crystalline structure. $TlInSe_2$ single crystal has a wide range of physical characteristics of practical importance, such as high photo- and roentgenosensitivity. The states localized in the band gap are responsible for most electronic processes occurring in semiconductors. The large anisotropy in chemical bonding (strong, ionic-covalent bonds within the chains and weak, van der Waals forces between the chains) enables effective doping of $TlInSe_2$ and $TlGaTe_2$ single crystals. The concentration and nature of dopants have a significant effect on the electrical properties of $TlInSe_2$ and $TlGaTe_2$ [1-3].

The $TlInSe_2$ and $TlGaTe_2$ compounds are structural analogs and crystallize in the tetragonal syngony. There is formation of broad regions of $(TlInSe_2)_{1-x}$ $(TlGaTe_2)_x$ solid solutions [1] in the $TlInSe_2$-$TlGaTe_2$ system. The roentgenographic properties [2], electrical conductivity [3] and thermoelectric properties of some $(TlInSe_2)_{1-x}$ $(TlGaTe_2)_x$ solid solution crystals are studied previously by us. Physical properties of solid solutions depending on the concentration and composition have been naturally changing.

The purpose of this work: thermodynamic study of the TlInSe$_2$-TlGaTe$_2$ system based on the data of physicochemical analysis, the study of electrical and dielectric properties of (TlInSe$_2$)$_{1-x}$ (TlGaTe$_2$)$_x$ solid solution crystals and influence of the composition on their dielectric characteristics.

THERMODYNAMIC DATA AND MODEL

Unlimited solid solutions are formed in binary and ternary systems, in the presence of the proximity of the crystal lattices of the initial components, ionic radii and electronegativity of anions. We carried out thermodynamic study of the TlInSe$_2$-TlGaTe$_2$ system based on the data of physicochemical analysis. Equation of temperature and concentration dependence of free energy ΔG_T^0 of formation of solid solutions of non-molecular compounds for ternary and quaternary systems [4, 5] was used as a model. Thermodynamic equation obtained for (TlInSe$_2$)$_{1-x}$ (TlGaTe$_2$)$_x$ solid solutions has the form (1).

$$\Delta G_T^0 (\text{kJ/mol}) = (1-x)\Delta H_{298.15}^0 (\text{TlInSe}_2) + x\Delta H_{298.15}^0 (\text{TlGaTe}_2) -$$
$$-(1-x)\cdot 10^{-3} T\Delta S_{298.15}^0 (\text{TlInSe}_2) - x\cdot 10^{-3} T\Delta S_{298.15}^0 (\text{TlGaTe}_2) +$$
$$+8.31\cdot 10^{-3} T[(1-x)\ln f(1-x) + x\ln x] - \tag{1}$$
$$-10^{-3} T\cdot \Delta C_{p,298.15}\cdot \left[\ln\left(\frac{T}{298.15}\right) + \left(\frac{298.15}{T}\right) - 1\right]$$

Enthalpy $\Delta H_{298.15}^0$ and entropy $\Delta S_{298.15}^0$ of formation of ternary compounds TlInSe$_2$ and TlGaTe$_2$ from binary compounds Tl$_2$Se, In$_2$Se$_3$ and Tl$_2$Te, Ga$_2$Te$_3$ in the solid state are used to calculate ΔG_T^0 of solid solutions from the composition (x). $\Delta H_{298.15}^0$ (TlInSe$_2$) = − 8.1 kJ/mol, $\Delta H_{298.15}^0$ (TlGaTe$_2$) = − 23.2 kJ/mol, $\Delta S_{298.15}^0$ (TlInSe$_2$) = − 0.1 J/ (mol · K), $\Delta S_{298.15}^0$ (TlGaTe$_2$) = − 3.1 J/ (mol · K). We obtained these thermodynamic functions of the compounds from the measurements of e.m.f. of concentration galvanic cells. Function $f(x) = x^3$ [6] is used to account for the configurational entropy of formation of non-molecular compound solutions in equation (1). $\Delta C_{p,298.15}$ Is the difference between the experimental values of the heat capacity of solid solutions (TlInSe$_2$)$_{1-x}$ (TlGaTe$_2$)$_x$ and TlInSe$_2$ and TlGaTe$_2$ compounds.

Thermodynamic values shown above have been put into equation (1). To calculate ΔG_T^0 of solid solutions of various compositions at different temperatures, we obtained expression (2)

$$\Delta G_T^0 (\text{kJ/mol}) = -8.08(1-x) - 15.07x + 0.09\cdot 10^{-3} T(1-x) + 3.10\cdot 10^{-3} T\cdot x +$$
$$+24.93\cdot 10^{-3} T[(1-x)\ln f(1-x) + x\ln x] - \tag{2}$$
$$-27x(1-x)^3 \cdot 10^{-3} T\left[\ln\left(\frac{T}{298.15}\right) + \left(\frac{298.15}{T}\right) - 1\right]$$

Equation (2) was used to calculate the phase diagram of TlInSe$_2$-TlGaTe$_2$ system. Concentration dependence ΔG_T^0 of solid solutions is characterized by a negative deviation from the additive dependence (figure 1). This agrees with the physico-chemical analysis and testifies the unlimited solubility in the solid state in TlInSe$_2$-TlGaTe$_2$ system.

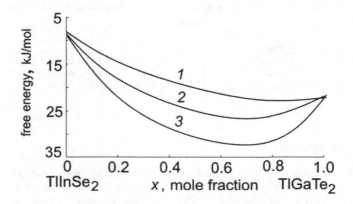

Figure 1. Dependence of free energy of formation of (TlInSe$_2$)$_{1-x}$ (TlGaTe$_2$)$_x$ solid solutions For temperatures: *1* – 300 K, *2* – 600 K, *3* - 900 K.

EXPERIMENTAL

The initial components TlInSe$_2$ and TlGaTe$_2$ are synthesized from pure chemical elements: thallium mark Tl-000, In-000, Ga-000, tellurium TB-3 and selenium OSCh-16-4. Synthesis of the samples was carried out by stoichiometric melting of weighed portions of TlInSe$_2$ and TlGaTe$_2$ components in evacuated to 10^{-3} Pa and sealed quartz ampoules [1-3]. High quality (TlInSe$_2$)$_{1-x}$(TlGaTe$_2$)$_x$ crystals were grown by Bridgman-Stockbarger method.

Individuality, the phase transition temperature, phase composition and homogeneity of the crystals were manipulated by DTA and X-ray phase analysis. Melting points of TlInSe$_2$ and TlGaTe$_2$ are 1040 K and 1048 K respectively.

Samples were prepared as powders for X-ray phase analysis. The diffractograms of the samples were recorded on X-ray diffractometer DRON-3 (Cu Kα-radiation) at room temperature. Roentgenographic data of the synthesized samples with coordination number 4 were indicated in a tetragonal type TlSe syngony (Table I).

Electrical and dielectric properties of the crystals were measured by the method described in [1-3]. Investigated samples formed flat capacitors. Ohmic contacts of samples are made by Ag paste. Measurements of the dielectric coefficients of studied single crystals were performed at fixed frequencies in the range 5×10^4–3.5×10^7 Hz by the resonant method using a TESLA BM 560 Qmeter. For electrical measurements, the samples were placed in a specially constructed screened cell.

Table I. Roentgenographic data on given compounds $TlInSe_2$, $TlGaTe_2$ and composition $(TlInSe_2)_{0.5}(TlGaTe_2)_{0.5}$ of solid solution with tetragonal type lattice and $D_{4h}^{18} - I4/mcm$ space group.

Phase	a, Å	c, Å
$TlInSe_2$	8.084 ± 0.002	6.844 ± 0.004
$TlGaTe_2$	8.430 ± 0.002	6.858 ± 0.004
$(TlInSe_2)_{0.5}(TlGaTe_2)_{0.5}$	8.280 ± 0.002	6.823 ± 0.004

RESULTS AND DISCUSSION

The frequency dispersion of real (ε') and imaginary (ε'') parts of complex dielectric permittivity, loss tangent (tan δ), and ac-conductivity (σ_{ac}) in the frequency range from 50 kHz to 35 MHz are studied in obtained crystals of $(TlInSe_2)_{1-x}(TlGaTe_2)_x$ solid solutions.

Figure 2 shows the frequency dependence of ε' for $(TlInSe_2)_{0.5}(TlGaTe_2)_{0.5}$ (curve 2) and $TlInSe_2$ (curve 1) for comparison. At frequencies $f \geq 10^5$ Hz ε' values of solid solution are 1.5-3.3 times higher than that of $TlInSe_2$.

The results demonstrate that the dielectric dispersion in the studied crystals $TlInSe_2$ and $(TlInSe_2)_{0.5}(TlGaTe_2)_{0.5}$ has a relaxation nature. ε' value of $TlInSe_2$ decreased by one order, and $(TlInSe_2)_{0.5}(TlGaTe_2)_{0.5}$ to 2.6 times with increasing frequency from 50 kHz to 35 MHz. In other words, dielectric permittivity dispersion of $TlInSe_2$ decreased significantly in comparison with $(TlInSe_2)_{0.5}(TlGaTe_2)_{0.5}$ solid solution.

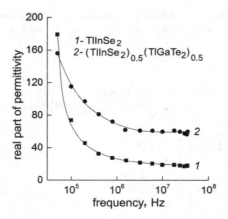

Figure 2. Frequency dispersion of real part of complex dielectric permittivity of TlInSe$_2$ single crystal (curve 1) and (TlInSe$_2$)$_{0.5}$(TlGaTe$_2$)$_{0.5}$ solid solution (curve 2). T = 300 K.

Same frequency dependence was observed for ε'' (figure 3). ε'' values for TlInSe$_2$ and (TlInSe$_2$)$_{0.5}$(TlGaTe$_2$)$_{0.5}$ were almost identical. ε'' values with increasing frequency up to 35 MHz decreased by more than two orders. In other words, ε'' in these crystals underwent a significant frequency dispersion.

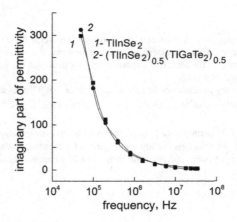

Figure 3. Frequency dependence of imaginary part of complex dielectric permittivity of TlInSe$_2$ single crystal (curve 1) and (TlInSe$_2$)$_{0.5}$(TlGaTe$_2$)$_{0.5}$ solid solution (curve 2). T = 300 K.

127

Frequency dependences of tan δ for both crystals are shown in figure 4. It is seen that tan δ of $(TlInSe_2)_{0.5}(TlGaTe_2)_{0.5}$ solid solution decreased compared to tan δ in $TlInSe_2$.

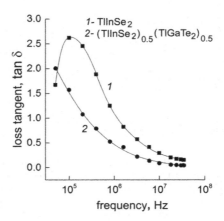

Figure 4. Dependence of tan δ on frequency in $TlInSe_2$ (curve 1) and $(TlInSe_2)_{0.5}(TlGaTe_2)_{0.5}$ solid solution (curve 2). $T = 300$ K.

The experimental frequency dependence of the dissipation factor for $(TlInSe_2)_{0.5}(TlGaTe_2)_{0.5}$ crystals is characterized with a monotonic descending with frequency (figure 4), which is evidence of the fact, that conductivity loss becomes the main dielectric loss mechanism at studied frequency range.

Behaviour of frequency dependence of ac-conductivity σ_{ac} of $TlInSe_2$ and $(TlInSe_2)_{0.5}(TlGaTe_2)_{0.5}$ solid solution in all studied frequency range (figure 5) is described by $\sigma_{ac} \sim f^{0.2}$ power law.

Dependence of high-frequency dielectric permittivity on the composition of $(TlInSe_2)_{0.5}(TlGaTe_2)_{0.5}$ solid solutions is shown in figure 6. Concentration dependence of ε' samples of $TlInSe_2$-$TlGaTe_2$ system indicates the formation of a continuous series of solid solutions.

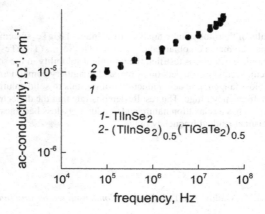

Figure 5. Frequency-dependent ac-conductivity of samples of TlInSe$_2$-TlGaTe$_2$ system at $T = 300$ K. 1 - TlInSe$_2$; 2 - (TlInSe$_2$)$_{0.5}$(TlGaTe$_2$)$_{0.5}$.

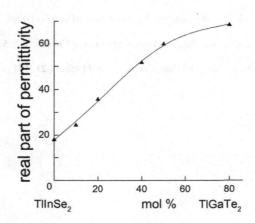

Figure 6. Dependence of dielectric permittivity on the composition of (TlInSe$_2$)$_{1-x}$ (TlGaTe$_2$)$_x$ solid solutions at $f = 35$ MHz.

CONCLUSIONS

We showed results of thermodynamic analysis of $TlInSe_2$–$TlGaTe_2$ system and high-frequency dielectric measurements on obtained crystals of $(TlInSe_2)_{1-x}$ $(TlGaTe_2)_x$ solid solutions. Thermodynamic calculations testified the unlimited solubility in the solid state in $TlInSe_2$-$TlGaTe_2$ system. Frequency dispersion of real and imaginary components of complex dielectric permittivity, loss tangent and ac-conductivity of crystals of solid solutions are studied in the 50 kHz-35 MHz frequency range. The results demonstrate that the dielectric dispersion in the studied solid solutions has a relaxation nature and conductivity loss becomes the main dielectric loss mechanism at studied frequency range.

REFERENCES

1. S.N. Mustafaeva, M.M. Asadov, A.I. Dzhabbarov, *Condensed matter and Interphases*, **15**, 150 (2013).

2. A. Sheleg, E. Zub, A. Yachkovskii, S.N. Mustafaeva, *Crystallogr. Rep.*, **57**, 283 (2012).

3. S.N. Mustafaeva, M.M. Asadov, A.I. Dzhabbarov, *Phys. Solid State*, **56**, 1055 (2014).

4. A. Mamedov, I. Mekhdiev, Z. Bagirov, *High Temperatures–High Pressures*, **29**, 689 (1997).

5. A.N. Mamedov, M.I. Zargarova, *Russian Journal of Physical Chemistry A*. **53**, 731 (1979).

6. M.M. Asadov, A.N. Mamedov, *Materials Chemistry and Physics*, **21**, 301 (1989).

Mater. Res. Soc. Symp. Proc. Vol. 1766 © 2015 Materials Research Society
DOI: 10.1557/opl.2015.420

Viscoelastic Behavior of Polymeric Optical Fiber

Alejandro Sánchez, Karla Y. Guerra, Andrés V. Porta, Susana Orozco*

Departamento de Física, Facultad de Ciencias, Universidad Nacional Autónoma de México,
Av. Universidad 3000, Col. Universidad Nacional Autónoma de México, C. U., Delegación
Coyoacán, C. P. 04510, D. F., México.
*E-mail: sos@ciencias.unam.mx

ABSTRACT

In this work, the viscoelastic behavior of a polymeric step-index optical fiber is studied,
and the loss factors η of their complex moduli are calculated. The loss factors of the Young
and shear moduli were determined from the measurement of the damping ratio γ of a simple
pendulum and a torsion pendulum respectively, using the Kelvin-Voigt model of the
viscoelastic theory. The shear and Young complex moduli can be used to study the optic-
viscoelastic behavior of a polymeric step-index optical fiber.

INTRODUCTION

The characterization of mechanical and thermal properties of optical fibers (OF) is
essential to determine their sensibility and operation range in their applications as sensors or
their stability in severe condition of optical transmission. On the other hand, the effect of
thermally and mechanically induced stresses on optical fibers have been extensively studied,
since the induced changes on the optical transmission can be used as a measure of an external
agent intensity in the sensor technology. Usually the thermal or mechanical behavior of the
OF is studied using the elastic theory [1]. Nevertheless, the polymeric optical fibers (POF)
have viscoelastic behavior rather than elastic [2]. The characteristics of POF based sensors,
being light weight, small in size, easily multiplexable, having high tensile strength and low
cost have attracted considerable attention in recent years [3], therefore it is very important to
determine the viscoelastic properties of these photonic materials.

A viscoelastic material is characterized by several elastic complex moduli; each one has
an accumulation term (storage modulus) and a dissipative term (loss modulus). The most
widely used methods for determining the complex moduli of elasticity and the loss factor
employ forced sinusoidal oscillations [4], and kinematic models based on springs and
dashpots (Voigt cells) are commonly used to study the viscoelastic behavior of polymeric
materials [2, 5, 6]. The advantage of this model consists of a minimum number of fitting
parameters and a reduced complexity of mathematical analysis. In this work, the Kelvin-
Voigt model is used to determine the shear and Young complex moduli of a polymeric step-
index optical fiber in the low frequency limit in which the moduli are independent of the
frequency [2]. Input parameters of the model are obtained experimentally from the
measurement of the damping ratio γ and the frequency in a torsion pendulum and a simple
pendulum. In these experiments, the optical fiber is subjected to dynamic loading through

low frequency oscillations, where the internal friction in the fiber resists the exciting stress. The characterization of viscoelastic properties and its effect on a strain state is essential to study the transmission of light in a POF under stress.

The viscoelastic materials have properties of an elastic solid and properties of a viscous fluid. If in addition, the mechanical properties are linear, the viscoelastic material can be modeled with a combination of springs (materials that behave according to Hooke law) and dampers (materials that behave like a fluid Navier-Stokes). In the Kelvin Voigt model, a spring and a dashpot in parallel characterized by the parameters E and μ_f respectively are considered as is shown in figure 1. When a periodic stress s with frequency ω is applied, a strain ε is produced in the material. A complex constitutive equation describes the behavior of the viscoelastic material,

$$\sigma^*(t) = (E' + iE'')\varepsilon^*(t). \tag{1}$$

Where the real and the imaginary terms are associated with the conservative and the dissipative parameters shown in Fig.1: $E' = E$ and $E'' = \mu_f \omega$. Since that the strain and the stress complexes differ in a phase we have that:

$$\sigma'(t) = E'\varepsilon(t) \text{ And,} \sigma''(t) = E''\varepsilon(t), \tag{2}$$

With ε the real strain, $\sigma'(t)$ the conservative stress, and $\sigma''(t)$ the dissipative stress. The modulus has the general form: $E^* = E' + iE''$, where E' is the elastic or storage modulus, and E'' is the damping or loss modulus; E^* is commonly expressed as $E^* = E'(1 + i\eta)$ where $\eta = \frac{E''}{E'}$ is the loss factor. The Eq. (2) can be applied to the shear stress and its strain in the complex shear modulus case.

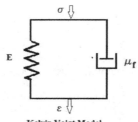

Kelvin Voigt Model

Figure 1. E is the conservative parameter and μ_f the dissipative one.

To determine the strain state of the POF, it is subjected to a dynamic load provided by an oscillating mass. The mechanical behavior of the optical fiber is associated with the dynamics of the mass attached at its end. Under ideal conditions the friction on the air and the upper support are negligible, but a dissipative force, originated in the POF, causes the damping characterized by the damping coefficient λ.

Shear modulus and loss factor

The optical fiber of length L, radius r, and cross section area A is loaded with a mass M. The torque balance for the pendulum is given by:

$$I\ddot{\theta} + \lambda\dot{\theta} + k\theta = 0, \tag{3}$$

Where I is the inertia moment of the pendulum, λ is the damping coefficient, and k is the torsion constant of the fiber. A known solution of the previous equation is $\theta = \theta_0 e^{-\gamma t}Cos(\omega t + \varphi)$, where the damping ratio $\gamma = \frac{\lambda}{2I}$, and the frequency ω satisfy the relation $\omega^2 = \omega_0^2 - \gamma^2$, and ω_0 is the natural frequency $\omega_0^2 = \frac{k}{I}$. The rotation angle θ is related to the shear strain angle α into OF as $\alpha L = \theta r$, therefore λ is the damping coefficient for α. Then the constitutive equation for the shear stress can be written as:

$$\sigma' = \frac{\tau'}{Ar} = \frac{k\theta}{Ar} = \frac{kL\alpha}{Ar^2} = \mu'\alpha, \tag{4}$$

$$\sigma'' = \frac{\tau''}{Ar} = \frac{\lambda\dot{\theta}}{Ar} = \frac{\lambda L\dot{\alpha}}{Ar^2} = \mu''\alpha. \tag{5}$$

Where σ' and μ' are the real shear stress and the real shear modulus, σ'' and, μ'' are the imaginary shear stress and the imaginary shear modulus. If we consider that $\alpha = Ce^{-\gamma t}e^{i\omega t}$, with C constant, $\frac{|\dot{\alpha}|}{|\alpha|} = \sqrt{\gamma^2 + \omega^2} = \sqrt{\gamma^2 + \omega_0^2 - \gamma^2} = \omega_0$, then the loss factor can be calculated from equations 4, and 5 as:

$$\eta = \frac{\mu''}{\mu'} = \frac{2\gamma}{\omega_0}. \tag{6}$$

Young modulus and loss factor

The motion equations of a mass M, in terms of the strain ε of a POF are given by:

$$ML_0(1+\varepsilon)\ddot{\theta} + \lambda_1 L_0(1+\varepsilon)\dot{\theta} + MgSin\theta = 0, \tag{7}$$

$$ML_0\ddot{\varepsilon} + \lambda_2 L_0\dot{\varepsilon} + kL_0\varepsilon = MgCos\theta. \tag{8}$$

Where L_0 is the unstrained fiber length, λ_1 and λ_2 are dissipative force constants, and k is the optical fiber stiffness constant. In the above equations, it was assumed that the dissipative force constant in the longitudinal and angular directions have different values. If in the equation 7 we consider that $\varepsilon \ll 1$, and $Sin\theta \approx \theta$, the solution $\theta = \theta_0 e^{-\gamma t}Cos(\omega t + \varphi)$, with $2\gamma = \frac{\lambda_1}{M}$, $\omega_{01}^2 = \frac{g}{L_0}$, and $\omega^2 = \omega_{01}^2 - \gamma^2$ can be assumed. On the other hand the equation 8 must be solved numerically.

The constitutive equation for the stress state on the fiber can be written as:

$$\sigma' = \frac{kL_0\varepsilon}{A} = E'\varepsilon, \tag{9}$$

$$\sigma'' = \frac{\lambda_2 L_0\dot{\varepsilon}}{A} = E''\varepsilon. \tag{10}$$

The loss factor can be obtained from equations 9 and 10,

$$\eta = \frac{E''}{E'} = \frac{\lambda_2\dot{\varepsilon}}{k\varepsilon} = \frac{2\gamma_2}{\omega_{02}^2}\frac{\dot{\varepsilon}}{\varepsilon} \approx \frac{2\gamma_2}{\omega_{02}}, \tag{11}$$

where $\omega_{02}^2 = \frac{k}{M}$.

EXPERIMENTAL

The damping ratios of oscillations of low frequency were used to determine the loss factor of the complex elastic moduli of a PMMA optical fiber. A good approximation is to consider that these kinds of optical fibers are isotropic materials [7]. For isotropic materials, experimental observations show [8] that the imaginary part of any complex stiffness is almost a constant proportion of the real part whatever the modulus is, which indicates that both Young or shear moduli could have loss factors of the same magnitude order. The frequencies were lesser than 10 Hz. In this excitation range the elastic moduli are independent of frequency, therefore the values obtained at low excitation frequency are also useful for static strain measurements; under these conditions an easy access display can be efficiently used. Torsion and simple pendulums of polymeric optical fiber with *PMMA* core and fluoropolymer coating with a diameter $d = 1 \pm 0.1\ mm$ were used. In both pendulums the fiber mass density was considerably less than the load mass density. On the other hand, a device with minimal friction was employed in the upper support.

In the torsion pendulum an optical fiber of $L = 75.0 \pm 0.1\ cm$ was loaded with an aluminum disc with mass $M = 259.5 \pm 0.1\ g$ and radius $d = 8.9 \pm 0.1\ cm$. The attenuation of the rotation angle was monitored with the video analysis software Tracker 4.81. The amplitude of rotation was observed marking a point on the disc.

In the simple pendulum, an iron mass with $M = 5.0 \pm 0.1\ g$ was hung on a fiber of unstrained length $L_0 = 45.0 \pm 0.1\ cm$. The damping in the amplitude of oscillation was monitored with the video analysis software Tracker 4.81 and the amplitude of oscillation as a function of the time $\theta(t)$ was obtained.

As expected, in both cases the curves correspond to exponentially decreasing functions of the time multiplied by sinusoidal functions which are in agreement with the solutions of equations 3 and 7. The loss factor of the shear and the Young moduli can be calculated from the damping ratios and the frequencies experimentally obtained, using the equations 6 and 11 respectively.

RESULTS AND DISCUSSION

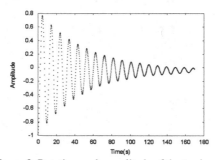

Figure 2. Rotation angle amplitude of the torsion pendulum.

134

Figure 2 shows the time variation of the rotation angle amplitude. The fitting of the experimental points gives the function $\theta(t) = e^{-\gamma t}Cos(\omega t + \pi)$ where $\gamma = 0.0209 \pm 0.0003 \ s^{-1}$, and $\omega = 0.663 \pm 0.013 \ s^{-1}$. The damping factor $\eta = 0.063 \pm 0.002$ was calculated from Eq (6). The real value of the shear modulus $\mu' = 1.24 \pm 0.19 \ GPa$ was measured previously, while in this work we have obtained the imaginary value of the shear modulus $\mu'' = 0.08 \pm 0.01 \ GPa$.

In figure 3 the time variation of the oscillation amplitude is shown. The function $\theta(t) = e^{-\gamma_1 t}Cos(\omega_1 t)$, with $\gamma_1 = 0.0050 \pm 0.0003 \ s^{-1}$ and $\omega_1 = 4.664 \pm 0.013 \ s^{-1}$, was obtained from the fitting of the experimental curve.

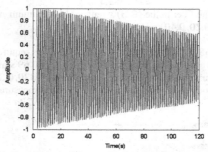

Figure 3. Time variation of the oscillation amplitude of the simple pendulum.

The previous values of γ_1 and ω_1 were introduced as parameters of the forced oscillation, while the damping ratio of longitudinal mode γ_2 was assumed as an adjustable parameter of the numerical solution of the equation 8. The time dependence of the strain along the fiber $\varepsilon = {\Delta L}/{L_0}$ is shown in figure 4. In the curve a longitudinal oscillation with $\omega = 9.328 \pm 0.013 \ s^{-1}$ is observed, the maximum strain ε is obtained in the equilibrium position of the mass M. The minimum strain varies with the damping of θ.

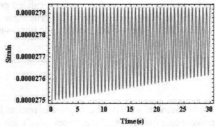

Figure 4. Time dependence of the strain ε along the fiber

In figure 5 the curve is shown in the $(0, 5) \ s$ range. Around the minimum and maximum amplitudes a second oscillation is also observed; this oscillation is clearly shown in the inset into the figure in the maximum around $(0.20, 0.40) \ s$. The frequency of this oscillation is $\omega_2 = 882.470 \ s^{-1}$. The corresponding damping ratio in this curve is $\gamma_2 = 0.005 \ s^{-1}$.

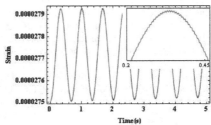

Figure 5. Strain ε along the fiber in the $(0, 5)$ s range. The maximum strain around $(0.20, 0.40)$ s is shown in the inset.

For a loss factor $\eta = 0.063$, a damping ratio $\gamma_2 = 26.470 \ s^{-1}$ is required. In equation 11, an oscillation with this γ_2 value is damped almost completely in $0.69 \ s$ as observed in figure 6. In this figure the maximum strain along the fiber around $(0.330, 0.350) \ s$ is shown. The real value of the Young modulus $E' = 3.030 \pm 0.054 \ GPa$ was measured previously; in this work we have obtained and concluded that $E'' = 0.180 \pm 0.003 \ GPa$.

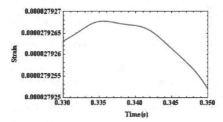

Figure 6. Maximum strain ε along the fiber around $(0.330, 0.350)$ s.

The intensity attenuation of a step-index optical fiber under localized pressure was investigated in a previous study using the elasto-optical theory [7]. The shear and Young complex modules calculated in this work will be used to evaluate the optic-viscoelastic behavior of the optical fiber under pressure.

CONCLUSIONS

We studied the viscoelastic behavior of a polymeric step-index optical fiber. The loss factors of the complex modules were calculated from the parameters of a damped oscillatory system using the Kelvin-Voigt model of the viscoelastic theory, considering the POF as an isotropic material. In the low frequencies limit, the Young modulus and shear modulus were determined, from measurement of the damping ratio γ, using a simple pendulum and a torsion one.

ACKNOWLEDGMENTS

This study is supported by PAPIIT: IN117014-3, DGAPA, Universidad Nacional Autónoma de México (UNAM). We appreciate the support from Posgrado en Ciencia e Ingeniería de Materiales, UNAM and Consejo Nacional de Ciencia y Tecnología (CONACYT)

REFERENCES

1. T.Z.N. Sokkar, M.A. Shams El-Din, A.S. El-Tawargy, *Optics and Lasers in Engineering*, **50**, 1223 (2012).

2. A. Stefani, S. Andresen, W. Yuan, O. Bang, *IEEE Sensors Journal*, **12**, 3047 (2012).

3. A. Kulkarni, J. Na, Y.J. Kim, S. Baik, T. Kim, *Optical Fiber Technology*, **15**, 131 (2009).

4. Brüel & Kjær Sound & Vibration Measurement A/S, *Measurement of the complex modulus of elasticity: A brief survey*. Application Notes, Denmark (1982).

5. M.A. Monclus, N.M. Jennett, *Philosophical Magazine*, **91**, 1308 (2011) DOI:10.1080/14786435.2010.504197.

6. C. Bernard, G. Delaizir, J.C. Sangleboeuf, V. Keryvin, P. Lucas, B. Bureau, X.H. Zhang, T. Rouxel, *Journal of the European Ceramic Society*, **27**, 3253 (2007).

7. A. Sánchez, S. Orozco, A.V. Porta, M.A. Ortiz, *Materials Chemistry and Physics*, **139**, 176 (2013).

8. B. Hosten, M. Castaings, *Composites*, **39**, 1054 (2008).

Mater. Res. Soc. Symp. Proc. Vol. 1766 © 2015 Materials Research Society
DOI: 10.1557/opl.2015.421

Analysis of the Mechanical Behavior of a Paperboard Profile

M. Rico[1*], J. M. Sandoval[1], L.A. Flores[1], N. Muñoz[1], P.A. Tamayo[1], R. G. González[2]

[1] Instituto Politécnico Nacional, Sección de Estudios de Posgrado e Investigación, Escuela Superior de Ingeniería Mecánica y Eléctrica, U. Azcapotzalco, Av. de las Granjas 682. Col. Sta. Catarina, Azcapotzalco, México D.F.
*Email: maryrico00@yahoo.com

[2] ESIQUIE-IPN, Laboratorio de Foto-Electrocatálisis, UPALM, C.P. 07738 México, D.F., México.

ABSTRACT

This paper shows the technological development for manufacturing corner angle sections or cardboard. Its manufacture involves splicing sheets (liners) of different weight Kraft papers joined with white glue. The thickness and strength of each profile is determined by the amount of spliced leaves and paperweight. There are two types of finishing in the profile, which are: natural finish Kraft wrapping paper and the white paper envelope. The second one is used to print images or logos on the exterior face for advertising purposes. They can withstand bending stresses for supporting buckling in horizontal and vertical position. These profiles are mainly used for packaging, protect corners, transportation and storage. A machine for manufacturing specialized linear process to obtain the required thickness is used. In this article, the basic load of an angular profile is analyzed by the finite element method using ANSYS 14 ®. Mechanical design considerations based on the mechanics of composite materials and the theory of laminated beams are considered. With the results of this analysis, load capacities like bending, buckling and deformation profiles are obtained. Furthermore, a comparison of three thicknesses of angular profiles supporting the mentioned loads is also presented.

INTRODUCTION

In recent decades, packaging systems evolved due to the high demand for products that companies require. A transport package is required to be strong and lightweight in order to be cost effective. Furthermore, it should be recycled because the environmental and economic concerns [1]. Despite of the undeniable appeal and importance of packaging, transportation and protection of products, *there are very little scientific research* works about the mechanical behavior of a paper board profile, its use in an optimum manner requires the knowledge of its mechanical behavior: elastic, inelastic, failure, etc., [2]. Due to the conditions of use required, in this paper, two important behaviors are analyzed, *bending* and *buckling*. Finite element simulation revealed that the profile has the ability to support the required loads. So, these data can be used as a tool to ensure the customer that the product will arrive in a better condition to the point of sale.

Paper board profile

The *paperboard profile* is a set of liners of Kraft paper or white paper of different weights joined with adhesive and forming an angle of 90°. Figure 1 shows the shape of the profile. The function is determined by its angular shape and the length of both sides can have different dimensions. Its strength and rigidity are determined by the amount of layers.

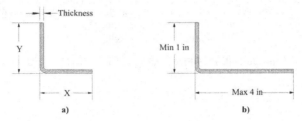

Figure 1. (a) Geometry of paperboard profile and (b) Geometry of asymmetrical paperboard.

Table I, shows the dimensions of the profile, the most common measure is the 2 x 2 inches.

Table I. Commercial profile (dimensions inches).

Side	Value				
X	1	1 ¾	2	3	4
Y	1	6	2	2	4

Figure 1-b shows that these profiles can be symmetrical or asymmetrical. They have a minimum size of 1 in. and a maximum size of 4 in. and the thicknesses range from 0.12 to 0.40 inches.

Mechanical Properties

The angular profile is considered as a *composite material* because the paper sheets are composed by two main components: the fiber and the matrix. With the proper combination of these components, a material with better properties is obtained. The fiber (paper) is the reinforcing component of the composite; this provides mechanical strength, stiffness and toughness [3]. The paper can be considered as an orthotropic elastic material, which has symmetry in the three perpendicular planes. If we take a piece of paper coming out of the roll, the longitudinal direction of the machine is known as MD (X), the transverse direction as CD (Y), and the thickness direction as ZD (Z) [4]. If we apply a uniaxial load to the sample in one of these three directions, we can make a small deformation as shown in Figure 2. The stress-strain ratio is defined as the elastic modulus or Young's modulus in the direction of the straining. One Young's modulus corresponds to each load case: E_{md}, E_{cd} and E_{zd}. In addition, for any of the three load modes, the Poisson ratio can be defined as the relation of the lateral contraction and the corresponding axial extension in the direction of the straining [3-5]. A matrix (white glue) is formed by an organic compound of high molecular weight, which makes chemical reaction with different compounds [6].

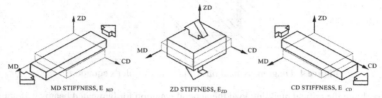

MD STIFFNESS, E $_{MD}$ ZD STIFFNESS, E$_{ZD}$ CD STIFFNESS, E $_{CD}$

Figure 2. Three modes of deformation in uniaxial tension.

Manufacturing process

In the flow diagram of Figure 3, the production process of this product is shown; this was carried out with a Kosebent Makinesi Machine K-3099 Model. The process begins with the selection of the paper; the paper can be composed of many layers. Once selected, the layers are uniformly glued and joined together at the same time. A "V" shape molding tool is used to bend the paper. Finally, the drying and cutting stages are followed. A total length of 6 meters can be achieved with this process.

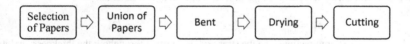

Figure 3. Flowchart production process.

Buckling analysis

According to these concepts, a composite material of a Kraft paper profile was analyzed. The test was carried on a set of 6 laminated liners of 90 lbs. Compression analysis for buckling and bending analysis that are based on the theory of laminated plates [3]. Figure 4 shows the direction of the applied load to an angular profile where L is the length of the sample.

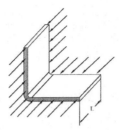

Figure 4. Diagram critical buckling load as Euler's equation.

To obtain the critical buckling load the general equation for laminated beams was applied and it is shown in equation 1 [7].

$$\varepsilon_1^0 = \frac{P}{bhE_0^1} \tag{1}$$

Where ε_1^0 corresponds to the longitudinal modulus, P is the load, b is the width of profile and h is the thickness. In this case, E_1^0 is equal to:

$$E_1^0 = \frac{\sum h^i E_1^i}{h} \tag{2}$$

Where h^i is the thickness of liner and E_1^i is the longitudinal modulus of the laminate. The displacement between the ends of the profile is obtained with Equation 3:

$$\delta_1^0 = \frac{PL}{bhE_1^0} \tag{3}$$

The laminate is unidirectional, so the stress σ_1^0 is constant throughout the thickness, and its value is:

$$\sigma_1^0 = \frac{P}{bh} \tag{4}$$

To obtain the critical buckling load of the beam it is necessary to start from the general equation for laminated beams:

$$E_1^f I \frac{d^4 \delta^0}{dx^4} + P(\delta^0) \frac{d^2 \delta^0}{dx^2} = 0 \tag{5}$$

Where:

$$\delta^0 = \delta_c^0 + \delta \tag{6}$$

$E_1^f =$ is the constant of bending. The sample was analyzed according to the Euler equation [8], under the principle of simply supported beam where the critical load is given by:

$$P_{cr} = \left(\frac{\pi}{L}\right)^2 E_1^f I = \left(\frac{\pi^2}{12}\right) \frac{E_1^f bh^3}{L^2} \tag{7}$$

Where L is the length of profile and I is the moment of inertia.

Bending analysis

The profile is a composite beam where the plane of the laminate is perpendicular to the direction of the load as shown in Figure 5. The beam was supported at the ends; different uniformly distributed loads as a pressure were applied from 100 to 400 N/m².

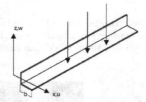

Figure 5. Direction of the applied load [3].

$$\varepsilon = \frac{-z\delta^2 w}{\delta x^2} = zk_x = \frac{zM_x}{J} \tag{8}$$

Similarly, the stress at the beam is equal to the strain in the beam multiplied by the modulus of elasticity in bending:

$$\sigma = \frac{-zE_f\delta^2 w}{\delta x^2} = \frac{zE_f M_x}{J} \tag{9}$$

The relationships of moments (M) and curvatures (k) are given by the flexion matrix or (D):

$$M_x = D_{11}k_x + D_{12}k_y \tag{10}$$

Thus: $\qquad\qquad J = E_f \tag{11}$

J is the product of the width of the profile.

Where:

$$E_f = \frac{1}{d_{11}} = \frac{12D_{11}}{h^3} \tag{12}$$

$$I = \frac{bh^3}{12} \tag{13}$$

EXPERIMENTAL

Material of the specimens

A set of 12 pieces of 2 x 2" specimens with 0.150" in thickness were used for the experiment. They were stored by 2 years at room temperature because that is the average storage time.

Procedure

Probes were manufactured by following the E8 ASTM standard for tension test. Their final shape is observed in Figure 6. For the bending test, a total of 4 specimens were manufactured. A Shimadzu universal tensile machine AG-IC 250 kN model was used to perform the tests as is shown in Figure 7.

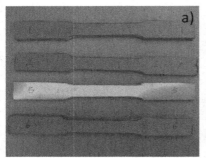

Figure 6. (a) Specimen of cardboard and (b) tensile test

Figure 7. (a) Shimadzu machine used by buckling test and (b) 3-point bending test.

Table II shows the results of the experimental tension test. An average load value of 190.787 kgf was obtained. The stress yield of the material is 1.38804 kgf/mm^2. Table III shows the results of the bending test. An average load value of 484.803 kgf was obtained and the elastic modulus was 136.791 kgf/mm^2. Table IV shows the results of the experimental flexion test. An average load value of 73.2423 kgf was obtained and the elastic modulus was 1058.59 kgf/mm^2.

Table II. Results of stress test.

Specimen	Max. Load (kgf)	Elastic Limit (kgf/mm²)
1	186.465	1.79189
2	170.571	1.53345
3	148.791	1.06498
4	257.319	1.16185

Table III. Results of buckling test.

Specimen	Max. Load (kgf)	Young's Modulus (kgf/mm²)
1	576.458	219.270
2	393.148	54.3119
3	269.428	144.136
4	-	-

Table IV. Results of bending test.

Specimen	Max. Load (kgf)	Young's Modulus (kgf/mm²)
1	88.5401	647.756
2	70.0497	1249.68
3	63.9235	1135.81
4	70.2423	1201.13

FEM simulation

Based on the finite element theory, the simulation solved by ANSYS® 14 is generated [9]. This method offers the advantage that the load-deformation response in any structures can be simulated for a variety of loads and for different material properties [4]. Shell elements were used to create the FEM model and to obtain the stress values [10-14], the element is composed by eight nodes with six degrees of freedom each one. Figure 8 show the stress and displacement analysis buckling.

The parameters shown in Table V are considered for the Baum test. The general dimensions of the specimen are 2" x 2"x 0.150" and a length of 21.65".

Table V. Elastic properties of linerboard 90 lb (GPa) [5].

E_x	E_y	E_z	V_{xy}	V_{xz}	V_{yz}	G_{yz}	G_{xz}	G_{xy}
7.46	3.01	0.029	0.117	0.109	0.021	0.104	0.129	1.80

After applying the loads, the maximum displacements were located at the center of the beam and the magnitude of the longitudinal stresses is observed in Table VI and VII respectively.

Results obtained with ANSYS® indicate that the maximum buckling stress is 6.1 MPa and the displacement is 5.057 mm and the results indicate that the maximum bending stress is 4.8 MPa and the displacement is 0.666 mm (Figure 9). Similar results were obtained between the experimental test and the numerical simulation (see Table VI and VII). These results give a numerical value suitable to understand the load levels that can be supported by this material which can also be used for the protection of products during packaging.

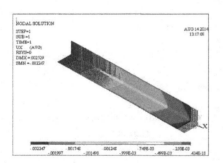 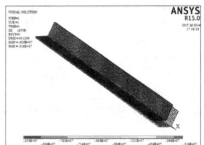

Figure 8. Stress and displacement analysis buckling.

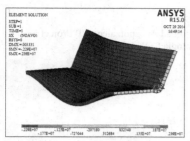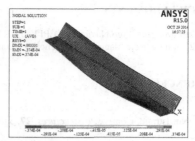

Figure 9. Stress and displacement analysis bending.

Table VI. Buckling results.

Load	Stress (MPa)	Displacement (mm)
100	1.52	1.266
200	3.05	2.529
300	4.58	3.793
400	6.10	5.057

Table VII. Bending results.

Load	Stress (MPa)	Displacement (mm)
100	1.2	2.529
200	2.4	4.592
300	3.6	6.655
400	4.8	0.066

CONCLUSIONS

The numerical and experimental results have a range of 29.23% with respect to the yield stress. Exceeded the expected supporting capacity of the profile, the analysis of the mechanical behavior suggests that there exist advantages and capabilities of using this type of profile. Also, suggest that selecting the appropriate profile avoids unnecessary expenses. On the other hand, numerical analysis simulation using ANSYS ® shows the mechanical behavior of the specimens, and this reduces cost, this implies less experiments required in the development phase of the process.

ACKNOWLEDGMENTS

Instituto Politécnico Nacional, SEPI ESIME Unidad Azcapotzalco, SIP Project No. 20140927 and Consejo Nacional de Ciencia y Tecnología CONACYT.

REFERENCES

1. T. Nordstrand, *Composite Structures*, **27**, 317 (1994).

2. Z. Aboura, *Composite Structures* **63**, 53 (2004).

3. A. Miravete, *Materiales Compuestos*, Tomo I. Edited by INO Reproducciones Zaragoza, 181 (2000).

4. J. Tryding, In-Plane Fracture of paper, Report TVSM-1008. Edited by Lund University, Division of Structural Mechanics, Sweden, 2 (1996).

5. G. Baum, *The elastic properties of paper: a review*, IPC Technical paper series **145** (1987).

6. L. Hollaway, Polymers Composites for Civil and Structural Engineering, Edited by Blackie Academic & Professional, (1993).

7. S. Timoshenko, Theory of plates and shells. 2nd Ed. McGraw-Hill Classic Text Book Reissue, 396 (1987).

8. H. Hahn, S. Tsai, *Journal of Composite Materials*, **7**, 102 (1973).

9. ANSYS Tutorial 2014, Shell 281.

10. J.L. Batoz, K.J. Bath, *Int J Numer Meth Eng.*, **15**, 812 (1980).

11. A. C. Gilchrist, J.C. Suhling, *Mech Cellulosic Mat*, **85**, 7 (1999).

12. T. Nordstrand, *Composite Structures*, **30**, 51 (1995).

13. A. Damatty, A. Mikhael, *Thin-Walled Struct*, **38**, 65 (2000).

14. J.L. Batoz, K. A. Bath, search for the optimum three-node triangular plate bending element. Edited by Rapport 82448-8, Massachusetts Institute of Technology, (1978).

Characterization of Materials Used in Coatings and Thin Films

Mater. Res. Soc. Symp. Proc. Vol. 1766 © 2015 Materials Research Society
DOI: 10.1557/opl.2015.422

Structural and Optical Characterization of ZnO Nanofilms Deposited by CBD-AμW

J. Díaz-Reyes[1*], R. S. Castillo-Ojeda [2], J. E. Flores-Mena [3] and J. Martínez-Juárez [4]

[1]Centro de Investigación en Biotecnología Aplicada, Instituto Politécnico Nacional. Ex–
Hacienda de San Juan Molino. Km. 1.5. Tepetitla, Tlaxcala. 90700. México.
[2]Universidad Politécnica de Pachuca. Km. 20, Rancho Luna, Ex-Hacienda de Santa Bárbara,
Municipio de Zempoala, Hidalgo. 43830. México.
[3]Facultad de Ciencias de la Electrónica, Benemérita Universidad Autónoma de Puebla. 18 Sur y
San Claudio S/N, Ciudad Universitaria. Col. San Manuel. Puebla, Puebla. 72570. México.
[4]CIDS-ICUAP, Benemérita Universidad Autónoma de Puebla. 14 Sur y San Claudio S/N, CU.
Edif. No. 137. Col. San Manuel. Puebla, Puebla. 72570. México.
* E-mail: joel_diaz_reyes@hotmail.com

ABSTRACT

ZnO was grown by Chemical Bath Deposition technique activated by microwaves (CBD-AμW) on corning glass substrates. The ZnO structural and optical properties are studied as a function of the urea concentration in the growth solution. ZnO chemical stoichiometry was determined by Energy-dispersive X-ray spectroscopy (EDS). The XRD analysis and Raman scattering reveal that ZnO deposited thin films showed hexagonal polycrystalline phase wurtzite type. The Raman spectra present four main peaks associated to the modes E_2^{high}, $(E_2^{high}-E_2^{low})$, E_2^{low} and an unidentified vibrational band observed at 444, 338, 104 and 78 cm^{-1}. The E_2^{low} mode involves mainly Zn atoms motion in the unit cell and the E_2^{high} mode is associated to oxygen motion. The observed emission peaks in the room temperature photoluminescence spectra are associated at vacancies of zinc and oxygen in the lattice.

INTRODUCTION

ZnO is an important semiconducting oxide because of it has a wide range of applications. ZnO can be used in piezoelectric devices or in optoelectronic applications, especially as a transparent electrode [1]. Its high electrical conductivity and optical transmittance in the visible region makes it useful for transparent conducting electrodes in flat panel displays or as optical windows in electroluminescent devices [2]. Among the various deposition techniques for ZnO thin films, DC reactive magnetron sputtering has received much attention, because of its flexibility, and because it offers good chemical composition control over extended areas.

Furthermore, the deposition technique offers the possibility to select the deposition rates in a wide range of values [3]. For more complex alloys, the stoichiometry of the films can be modified by changing the substrate temperature, the pressure and the reactive atmosphere used during the deposition process. Moreover the properties of the films depend too on the sputtering power and post annealing processes on the films [3].

In this work reports the growth and the structural and optical characterization of ZnO nanofilms deposited by chemical bath deposition technique activated with microwaves (CBD-AµW) on glass corning at various urea ratios. Effects of urea ratios on crystalline quality, grain size and surface morphology of the thin films were studied by X-ray diffraction, SEM-EDS, Raman spectroscopy and photoluminescence.

EXPERIMENTAL

As result of the chemical reactions in the CBD-AµW the ZnO is obtained and deposited in thin film form on corning glass substrates. The solutions of the precursor reagents are prepared at 300 K using deionized water of 18.2 MΩ-cm of resistivity with the purpose of diminishing the residual impurity concentration in the grown material. The molar fractions used for deposition were as follows: $[Zn (NO_3)_2] = 0.1$ M that was maintained constant while the urea molar fractions were varied from 0.1, 0.2, 1.0 M. The substrates were submerged in the solution and NH_4OH was added to it. Afterwards, they were undergoing microwave irradiation during 5.0 min at maximum power keeping the temperature constant and after irradiating them at lowest power for 40 min of a microwave oven. Finally, they were rinsed with strong agitation and were dried with gaseous nitrogen. Structural characterization of the ZnO samples was carried out by means of X-ray diffraction (XRD) in a Bruker D8 Discover diffractometer, parallel beam geometry and monochromator of gobel mirror, CuKα radiation, 1.5406 Å, in the range of $20° \leq 2\theta \leq 80°$. The XRD data were refined using the programs POWDERX and DICVOL04 to determine the crystalline system, the parameters of unit cell.

The compound stoichiometry was obtained by measurements of energy dispersive spectroscopy (EDS) that were carried out in an LEO 438VP system, with W.D. of 26 mm using a pressure of 20 Pa. Raman scattering experiments were performed at 300 K using the 6328 Å line of a He-Ne laser at normal incidence for excitation. The light was focused to a diameter of 6 µm at the sample using a 50x microscope objective. The nominal laser power used in these measurements was 20 mW. Scattered light was analysed using a micro-Raman system (Lambram model of Dilor), a holographic notch filter made by Kaiser Optical System, Inc. (model superNotch-Plus), a 256×1024-pixel CCD used as detector cooled to 140 K using liquid nitrogen, and two interchangeable gratings (600 and 1800 g/mm).

Photoluminescence was taken with a solid state laser at 325 nm with 60mW as excitation source and a Sciencetech 9040 monochromator was used to perform the sweep of wavelength at room and low temperature in a cryostat Cryogenics measured at room temperature.

RESULTS AND DISCUSSION

The chemical composition of the obtained ZnO thin films was carried out measurements of energy dispersive spectroscopy (EDS) to some typical samples with urea ratios: 1:1, 1:5 and 1:10. The results of such measurements are shown in Table I, in which are included the atomic and mass percentages. From these results is observed that the urea ratio 1:5 is the that one gives a best chemical stoichiometry and starting from it at a higher urea concentration in the growth solution gives a greater presence of zinc in the grown material and an absence of oxygen. Considering that each unit cell contains two zinc atoms and two oxygen atoms, the atomic weight of the ideal unit cell is ~162.74 corresponding to 19.66% oxygen atoms and 80.33% zinc atoms, then when a stoichiometric deviation of ideal unit cell occurs it could establish a correspondence between vacancies or interstices of some of the compound elements (V_O, V_{Zn}, Zn_i, O_i).

Figure 1 shows the diffractograms of the ZnO films; the pattern 36-1451 is enclosed in the inferior part of graphs, which indicate that in all the zinc nitrate and urea ratios the synthesized material is ZnO and besides there is not a change of crystalline phase. ZnO has hexagonal polycrystalline wurtzite type structure and the variation in the solutions concentrations is not remarkably observed in the diffractograms. From the X-ray patterns were not possible to determine the influence of the urea concentration in the growing direction of the films, as is clearly observed in Fig. 1.

The labelled indices show that all the samples have the ZnO wurtzite-type structure and no evidence of urea-related secondary phases, within the detection limit of the instrument. ZnO commonly crystallizes in a hexagonal wurzite type structure in the most of the synthesizing methods, with lattice parameters of unit cell of $a = 3.2495$ Å and $c = 5.2069$ Å [4]. By refinement of X-ray experimental data finds the unit cell parameters for all studied urea concentrations, whose parameter average values of the ZnO thin films are $a = 3.253$ Å and $c = 5.212$ Å, with a tendency to increase in both parameters and a similar tendency to increase or decrease among their values starting from the molar relation 1:4 to 1:10. However, the tendency of lattice constant values in the molar relations 1:1, 1:2 and 1:3 is opposed, which can be explained if it is considered the stress when the film is deposited, this is valid for the three molar ratios.

For the molar relation 1:4 the film is deposited with compressive stress and starting from the relation 1:5 again it is deposited with strain. In order to obtain more structural information, the mean grain size of the deposited films was evaluated using the Scherrer equation. The mean grain size was calculated from hexagonal (101) reflection of ZnO for the all samples, obtaining an average grain size was about 26 nm, implying that under these experimental conditions cannot be obtained nanoparticles with strong quantum confinement.

Table I. Results of the analysis by EDS of the samples with the relations 1:1, 1:5 and 1:10.

Ratio [Zinc]/[Urea]	O (At. %)	Zn (At. %)	O (wt. %)	Zn (wt. %)
1:10	47.01	52.99	17.83	82.17
1:5	50.08	49.92	19.71	80.29
1:1	52.43	47.57	21.24	78.76

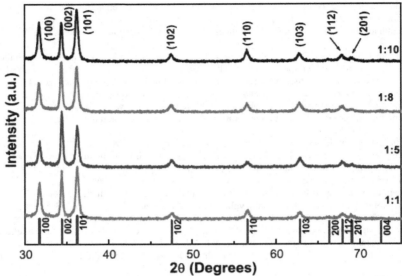

Figure 1. X-ray patterns of the ZnO films synthesized by CBD-AµW, where is shown that there is no change of crystalline phase.

Assuming that the volume of the unit cell remains around 47.78 Å^3 for all cases and because of the atomic weight of the zinc atom is four times greater that of oxygen atom, it could be expected that a great presence of zinc atoms in the structure should correspond to a higher density and the contrary should happen for a greater presence of oxygen atoms. However, when the EDS results are analysed considering the densities can see that for a greater Zn concentration in the material the density diminishes, which indicates that the zinc is present interstitially in the structure in such a way that relaxes the chemical bonds. From Table 1 is observed that the sample with ratio 1:10 contains higher zinc concentration, while the sample with ratio 1:1 contains more zinc vacancies.

From Table I observes that the sample with ratio 1:10 contains higher zinc concentration, while the sample with ratio 1:1 contains zinc vacancies. Relating these results with Figures 2a, 2b and 2c obtained by scanning electronic microscopy (SEM) on the samples 1:1, 1:5 and 1:10, respectively. It is noticed that the presence of zinc in the structure or urea quantity in the solution retard the growth of the structures although the surface morphology remains. Figure 2 shows the SEM images of ZnO films that were homogenized and agglomerated with particles of several sizes dependent of the ratio (nitrate: urea).

Figure 2. SEM images of ZnO structures samples with molar ratios; (a) 1:10, (b) 1:5, (c) 1:1.

Figure 3 shows polarized Raman spectrum of the ZnO film with ratio 1:4 measured in backscattering configuration, which allows the observation of all active modes and besides some ones are well-resolved for the deposited films studied in this work. The inset shows the Raman spectra, which show a similar behaviour and only variations in the intensity of the bands. The sample 1:4 that is closely stoichiometric presents the more intense Raman spectrum and is better resolved and as the urea concentration is increased/diminished in the solution the spectra decrease though they are nearly similar.

The Raman spectra are dominated by two intense modes: sharp E_2 modes sited at 104 and 444 cm^{-1}. The low-frequency E_2 mode, involving mainly Zn atoms motion in the unit cell, displays an extremely narrow linewidth. The E_2^{high} mode displays a clear asymmetry toward low frequencies that we shall discus in more detail below and it is associated to oxygen motion. The Raman spectrum can be deconvoluted in Lorentzian line shape signals through a standard fitting procedure performed in the whole studied interval and the results are plotted using solid lines, which are shown in Figure 3 for the sample with urea ratio 1:4. Reasonable fits are achieved with Lorentzian curves, whose frequencies are shown in figure and which allow to identify the vibrational bands present in the Raman spectrum. In contrast with previous works, where the $A_1(LO)$ mode could not be detected for excitation wavelengths longer than 406.7 nm [5], we detect a weak $A_1(LO)$ mode at 578 cm^{-1}. In addition to the strong E_2 mode a new peak appears at 388 cm^{-1}, which can be assigned to the $A_1(TO)$ mode.

We detect a strong mode, a low frequency one at 104 cm^{-1}, which could be associated to the E_2^{low} mode; this is related at zinc atoms motion in lattice. The $2E_2^{high}$ mode can be observed at 195 cm^{-1} in the Raman spectrum. Finally, there are two intense modes at 78 and 115 cm^{-1} that could not be identified. One of the peaks in the intermediate-low-frequency region is observed at 338 cm^{-1}. This mode had been previously assigned to transverse acoustic overtone scattering at M [5-7]. However, studies made by Cuscó of temperature dependence of the Raman intensity clearly indicate that this one is a difference mode, see Figure 3. The frequency of this mode is in good agreement with the difference between the E_2^{high} and E_2^{low} frequencies measured in our samples (340 cm^{-1}) [8]. The E_2^{high} - E_2^{low} difference contains symmetries A_1, (E_1, E_2). According to the calculated phonon dispersion relations [7], the 338 cm^{-1} mode could also contain contributions from $[TO-TA]_{A,L,H}$ differences. Another prominent peak is observed at 195 cm^{-1}, which exhibits A_1 symmetry and thus can be attributed to a TO overtone [8]. The peak at 534 cm^{-1} is of A_1 symmetry and can be attributed to $2B_1^{low}$ and LA overtones along L-M and H. It is noted that the E_2^{high} mode exhibits a visibly asymmetric line shape with a low-frequency tail, which is associated to O motion in the lattice. This is quite apparent from Fig. 5 when one compares the line shapes of the E_2^{high} and $A_1(TO)$ modes. The asymmetry of the E_2^{high} mode cannot be ascribed to lattice disorder. Furthermore, isotopic broadening is negligible for the E_2^{high} mode since it mainly involves O motion, and O is nearly isotropically pure. The line-shape broadening is then mostly determined by anharmonic phonon-phonon interactions. These can result in strongly distorted peaks when resonant interaction with a band of second-order combinations takes place (Fermi resonance), as is the case, for instance, for the GaP TO mode [9], where the presence of van Hove-type singularities in the density of states (DOS) of the TA+LA combination band gives rise to a highly asymmetric TO mode, which develops a side band at high pressure [9]. A similar situation occurs for the E_2^{high} mode, as its frequency lies close to a ridge-like structure of the two-phonon DOS corresponding to TA+LA combinations in the vicinity of the K point [10]. Whereas wavevector conservation restricts the phonons involved in first-order Raman scattering to those with k~0, phonons from the entire Brillouin zone take part in second order Raman scattering. Therefore, second-order spectra usually display feature-rich structures, which are determined, on the one hand, by the phonon DOS and, on the other hand, by the selection rules of the two-phonon scattering processes. Selection rules for two-phonon Raman scattering in crystals with the wurtzite structure were reported by Siegle [6].

According to density-functional theory calculations [7] the phonon DOS of ZnO presents a frequency gap between acoustic and optical modes that extends from 270 to 410 cm⁻¹. The second-order spectra may then be divided into three regions: (i) the low-frequency region (approximately 160–540cm⁻¹) dominated by acoustic overtones, (ii) the high-frequency region (820–1120cm⁻¹) formed by optical overtones and combinations, and (iii) the intermediate-frequency region (540–820cm⁻¹) where optical and acoustic phonon combinations occur.

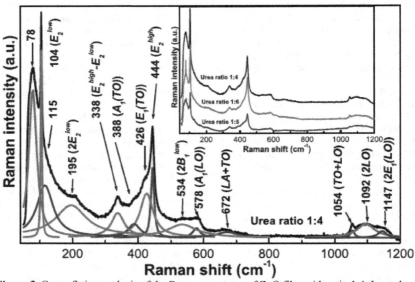

Figure 3. Curve fitting analysis of the Raman spectrum of ZnO film with ratio 1:4. Inset shows the Raman spectra of ZnO films for molar ratios 1:4, 1:5 and 1:6.

The second order features are labeled with their respective frequencies on the room temperature Raman spectrum. The most prominent second-order features occur in the high-frequency region and correspond to LO overtones and combinations involving LO modes. The broad peak at 1148 cm^{-1}, contains contributions of $2A_1(LO)$ and $2E_1(LO)$ modes at the Γ point of the Brillouin zone, and possibly also of $2LO$ scattering by mixed modes from the rather bands along the A-L-M-line. The vibrational peak at 1093 cm^{-1} can be attributed to $2LO$ at the H and K points of the Brillouin zone. Finally a thin peak can be observed in the RT Raman spectrum at 1052 cm^{-1}, which we assign the mode to $TO+LO$ combinations at the A and H points.

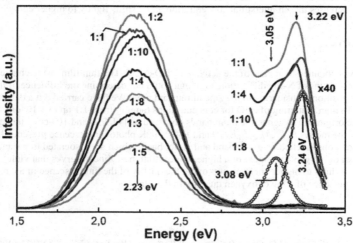

Figure 4. RT-PL spectra of ZnO films deposited by CBD-AµW for the different urea ratios. It shows the deconvolution of PL spectrum of sample with ratio 1:8.

The PL spectra of some ZnO films are shown in Figure 4. The PL spectra presented a dominant radiative band at 2.22 eV and a weaker band around 3.22 eV. Assignation of each transition is accomplished studying the behaviour of the PL spectra as function of the urea concentration. The energy positions of peaks have been determined by a quantitative fit to the experimental PL spectra using a sum of Gaussian line distributions, the dominant peaks were fit first and the additional peaks were added as necessary. The quantitative fit of the experimental PL spectra, using the sum of Gaussian curves, suggests that the PL emission consists of three emission bands with photon energy centred at 2.23, 3.05 and 3.22 eV that are associated to green-yellow, violet and ultraviolet bands. The ultraviolet and violet luminescence observed in the case has been attributed exciton-related near–band-edge luminescence while the band sited around 2.23 eV is commonly referred to as a deep-level or trap emission [11]. The energy level corresponding to Zn interstitials lies just below the conduction band, and it can trap photoexcited electron followed by their radiative recombination with holes in the valence band. From Table I that shows the EDS results where are presented the atomic percent of zinc and oxygen and the corresponding mass percent of the same atoms. From these data is found that in the first molar ratio (1:1) there is a zinc deficiency where the most probably defect present is the oxygen interstitial, as can see in Figure 3 this sample has an excitonic signal more intense; in the second ratio (1:5) the film composition is more stoichiometric and the defects concentration is probably more lower, in last presented ratio (1:10) the composition has an oxygen deficiency and the zinc interstitial is predominant defect. As mentioned above, these have been associated with the radiative decay of intrinsic defects bound to neutral vacancies associated to O and Zn [12]. As is observed in Figure 4, band ultraviolet increases as the Zn interstitials are predominant and shifts lightly toward higher energies, about 20 meV. The green band is intense and dominant in the stoichiometric/nonstoichiometric samples and it does not shift appreciably. The violet band increases to increases the Zn interstitial concentration and it shifts lightly to higher energies.

CONCLUSIONS

ZnO was obtained by means of the activated CBD-AμW. The thin films were characterized structurally by means of X-ray diffraction, although could not determine the influence of the solution composition, considering to the zinc nitrate and urea. We have carried out a detailed study of the Raman scattering of ZnO for concentrations ranging from 1:1 up to 1.10, one has discussed the origin and assigned the main bands observed at 444, 338 and 104 cm^{-1} that are associated at the modes E_2^{high}, $(E_2^{high}-E_2^{low})$ and E_2^{low}. 300K photoluminescence presents three radiative transitions labelled by green band and violet band, which are associated to vacancies and interstices of oxygen and zinc. For a higher concentration of Zinc observes that violet emission band shifts to higher energies improving the quality of the luminescence in the region UV with the presence of a lower oxygen quantity in the film.

REFERENCES

1. S.Y. Lee, E. S. Shim, H.S. Kang, S.S. Pang, J. S. Kang, *Thin Solid Films*, **437**, 31 (2005).
2. X.T. Hao, T.L. Wei, K.S. Ong, F. Zhu, *J. Cryst. Growth*, **287**, 44 (2006).
3. K. Ellmer, *J. Phys. D: Appl. Phys.*, **33**, R17 (2000).
4. C. Jagadish, *Zinc Oxide Bulk, Thin Films and Nanostructures*, First edition, Ed. Elsevier, England (2006).
5. J. M. Calleja, *Phys. Rev. B*, **16**, 3753 (1977).
6. H. Siegle, G. Kaczmarczyk, L. Filippidis, A.P. Litvinchuk, A. Hoffmann, C. Thomsen, *Phys. Rev. B*, **55**, 7000 (1997).
7. J. Serrano, A.H. Romero, F.J. Manjón, R. Lauck, M. Cardona, A. Rubio, *Phys. Rev. B*, **69**, 1 (2004).
8. R. Cuscó, E. Alarcón Lladó, J. Ibáñez, L. Artús, J. Jiménez, B. Wang, M.J. Callahan, *Phys. Rev. B*, **75**, 1 (2007).
9. B.A. Weinstein, *Solid State Commun*, **20**, 999 (1976).
10. J. Serrano, F.J. Manjon, A.H. Romero, F. Widulle, R. Lauck, M. Cardona, *Phys. Rev. Lett.*, 90 (2003).
11. B. Lin, Z. Fu, Y. Jia, *Appl. Phys. Lett.*, **79**, 943 (2001).
12. Y. Sato, S. Sati, *Thin Solid Films*, 281 (1996).

Mater. Res. Soc. Symp. Proc. Vol. 1766 © 2015 Materials Research Society
DOI: 10.1557/opl.2015.423

Study of Corrosion Behavior of Polyuretane/nanoHidroxiapatite Hybrid Coating in Hank Solution at 25 °C

G. Carbajal-De La Torre, A.B. Martinez-Valencia, A. Sanchez-Castillo, M. Villagomez-Galindo, M.A. Espinosa-Medina*

Facultad de Ingeniería Mecánica, Universidad Michoacana de San Nicolás de Hidalgo, C.P. 58000, Morelia, Michoacan México.
* E-mail: marespmed@gmail.com

ABSTRACT

The study of corrosion behavior of polyurethane/nanohydroxyapatite hybrid coating in aerated Hank solution at 25 °C by Potentiodinamic and Electrochemical Impedance techniques was realized. The nanohydroxyapatite (nHA) powders were synthesized by ultrasonic assisted co-precipitation wet chemical method and then mixed with pure polyurethane (PU) during the polymerization. Results were supported by SEM morphologic characterization. Results showed that good corrosion resistance of hybrid coating, showing small corrosion product layer formation. Corrosion mechanisms are affected by an increasing of polarization resistance, promoting decreasing in the corrosion rates. Diffusion of ionic species was the governing mechanism in the corrosion behavior of polyurethane/nanohydroxyapatite hybrid coating.

INTRODUCTION

The needing on an alternative solution in the use of biomaterials have promoted the researching on basic science and technology about those [1,2]. Metallic implants are applied as temporally assist in healing implements as plates and screws due high corrosion resistance and enough biocompatibility [1], however corrosion products might be in high concentrations causing allergic reactions by toxicity. Some metallic alloys as the stainless steels, platinum and Ti-6Al-4V alloys have improved high corrosion resistance [3]; however, those could promote toxic ion release [1]. A way to mitigate higher corrosion rates in the body solution [3, 4] is to apply a protective coating to avoid direct contact with the medium [5-10].

In this sense, ceramic/polymer biomaterials have been considered as potential application as those. Composite materials based in biodegradable and biocompatible synthetic polymers reinforced with hydroxyapatite, which can be synthesized easily, has a higher potential for several tissue regeneration processes due its excellent biocompatibility and its mineralogy similitude [10]. The polyurethane (PU) with hydroxyapatite (HAp) based biocomposites have been presented excellent results [9]. In this work, electrochemical techniques were used to study the corrosion behavior of PU/nHA composite coatings in the simulated Hank solution at 25 °C. Results were complemented by morphology and chemical characterization by scanning electron microscopy (SEM).

EXPERIMENTAL

The starting nHA powders were synthesized by ultrasonic assisted co-precipitation wet chemical method as described previously [8]. Then, obtained nHA powders were mixed with pure PU during the polymerization [8, 10] at different weight ratios: 5, 10 and 20 wt. %.nHP. AISI-1018 steel sheets were used as substrate to apply the hybrid coatings by dip-coating technique [11, 12]. Prior to coating applying, sheets surfaces were grinded with Emery papers up to 600 grind size, washed and cleaned with acetone and dry air.

Substrate samples were coated by dip-coating method as single layer coatings at constant withdrawal speed of 20 cm/min then dried at 60°C for 24 h. Coatings immersed in Hank solution at 25 °C were evaluated by direct and alternate current electrochemical techniques. Electrochemical tests were realized by using a three electrode cell array with a potentiostat Gill ACM Instruments connected to a computer. A graphite bar was the auxiliary electrode; a saturated calomel electrode was the reference, and coated sample with an exposed area of 0.36 cm^2 was work electrode.

Potentiodynamic tests were carried out from -650 to 1800 mV volts at the scan rate of 1 mV/s. LPR data was obtained each 15 min during 24 hours at the scanning rate of 1 mV/s. Electrochemical Impedance Spectroscopy (EIS) measurements were carried out at corrosion potential within the frequency range of 6 x10^4 to 1 x10^{-1} Hz, applying a sinusoidal signal of 20 mV in amplitude. Coating samples were morphology characterized by scanning electron microscopy (SEM).

RESULTS AND DISCUSSION

Previous results of the characterization of hybrid composites by XRD and FTIR [9] used to apply the coatings evaluated in this work, presented XRD results showing the characteristic pattern of crystalline hydroxyapatite, but secondary phases were not present; also PU pattern was present in all samples. The IR spectra for PU, nHA powder and PU/nHA for the coatings were corroborated because of all characteristic bands were observed [9].

Figure 1 shows PU and PU/nHA coatings morphologies by SEM before electrochemical tests. Those coatings showed homogeneous nHA particle distribution within PU, in addition there were no visible areas of particle agglomerations found. Homogeneous porosity distribution of PU and composites was observed, showing sizes within the order of 1 to 5 μm.

Potentiodynamic test results presented in figure 2, show that PU/nHA wt.10% coating (plot d) followed by PU coating (plot b) have higher protective behavior into the Hank solution at 25 °C. Both coating samples showed more positive E_{corr} values (open circuit potential; OCP) around of -348 and -339 mV vs SCE respectively. They more positive E_{corr} potentials, also the lower anodic current density value observed, were associated to lower electrochemical activity of coatings than the coatings with 5 and 20 wt.% HP concentrations (plots c and f). In specific, PU/nHA wt.10% coating showed an i_{corr} (1.11 x10^{-5} mA/cm^2) decreasing in more than one order of magnitude presenting an anodic current density decay associated to semipassive or higher resistive behavior [12]. Electrochemical parameters obtained of potentiodynamic results are showed in Table I.

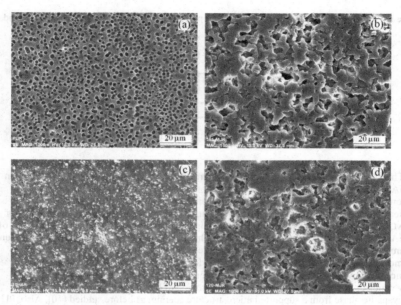

Figure 1. Images by SEM of: a) PU, b) PU/nHA wt.5%, c) PU/nHA wt.10%, y d) PU/ nHA wt.20% hybrid coatings before electrochemical test.

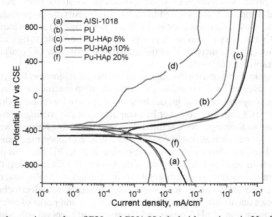

Figure 2. Potentiodynamic results of PU and PU/nHA hybrid coatings in Hank solution at 25 °C.

Table I. Electrochemical parameters obtained of potentiodynamic results by intercept method.

Sample	E_{corr} (mV vs SCE)	I_{corr} (mA/cm^2)	βa (mV)	βc (mV)
AISI-1018	-457	2.3×10^{-4}	86	104
PU	-339	1.1×10^{-4}	60	14
PU/nHA wt.5%	-385	2.0×10^{-4}	34	20
PU/nHA wt.10%	-348	1.1×10^{-5}	113	45
PU/nHA wt.20%	-444	2.6×10^{-3}	55	80

The PU/nHA wt.5% coating showed a -385 mV vs SCE value of E_{corr} slightly lower than PU/nHA wt.10% coating, and an i_{corr} value similar to showed by PU coating. However, the incorporation of wt.20% of nHA in the PU did not improved protective behavior over the metallic substrate (fig. 2). Therefore, the nHA addition to PU improves the protective behavior up to wt.10% of nHA, but was not apparent with higher concentration. Electrochemical result of PU/nHA wt.20% coating was associated to the higher path density (fig. 1d) due the interrelation of ceramic phase (nHA) with the PU matrix. Thus electrolyte solution cans permeate easily to reach metallic interface. Although, nHA wt.20% addition does not improved corrosion resistance, it showed a very important result for biomedical applications, due the higher path density which as microporous structure in the improving of bioactivity and grooving of hydroxyapatite phase from a supported microstructure coating, as before studied [10]. Also, PU biodegradation can be accelerated due more volume electrolyte permeation into the microporous coating structure.

In accord to above, corrosion mechanisms of coatings in the Hank solution are important to be established. EIS results show the nHA wt % addition effect on the corrosion behavior into the solution, presented in figure 3. Addition of nHA wt % increases the magnitude of impedance magnitude (Z) up to nHA wt.10% concentration (Z Bode plots, fig. 3c) and increases the range o frequencies of capacitive activity of corrosion processes (phase bode plots, fig. 3c). In similar way, in the Nyquist plot (fig. 3a) increasing in the impedance resistance (Z' real) and capacitive magnitude (Z" imaginary) are observed.

Coating applying on the metallic substrate modifies the corrosion mechanism; the depressed semicircle presented by the metallic substrate, shows a mechanism of activation affected by the surface roughness (that at higher frequencies) mixed with finite diffusion behavior at low frequencies (fig. 3b); however EIS responses for the coatings show a increasing in the resistive mechanism forming a wide semicircle, but at the lower frequencies a characteristic inductive behavior was observed (fig. 3a). Thus, corrosion mechanisms presented by the PU/nHA hybrid coatings have a resistive component due the PU phase in parallel with a finite diffusion mechanism observed in the range of medium to low frequencies; that associated to transport mechanism of oxidant species thought coating thickness. At the lower frequencies an inductive behavior was presented, which was associated to intermediate electrochemical reactions or ionic species adsorption occurred between the paths and pores of microstructure due the physicochemical interaction of nHA phase and the simulated corporal solution. The interaction of Hank solution with the PU and PU/nHA coating interfaces after 24 hours of immersion are showed in figure 4.

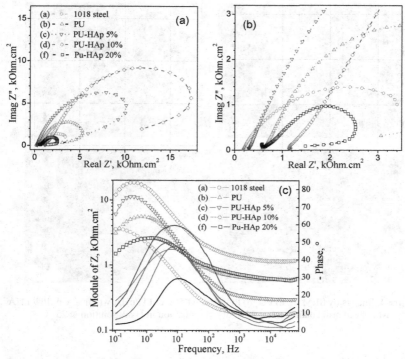

Figure 3. EIS results of PU and PU/nHA hybrid coatings in the Hank solution at 25 °C. a) Nyquist plot, b) zoom of area in Nyquist plot and c) Z magnitude and phase Bode plots.

As the potentiodynamic, the EIS results show that higher impedance resistance was obtained by the PU/nHA wt.10% hybrid coating, followed by the PU/nHA wt.5% and PU coatings. In that way, SEM images show the surface morphology of corroded samples after 24 hours of immersion at OCP conditions (fig. 4); where the PU/nHA wt.10% hybrid coating (fig. 4c) showed lower corrosion products on the coating surface and low corrosion dissolution and pore density, improving higher corrosion resistance, as observed in electrochemical evaluations (fig. 2 and fig. 3). In contrast, the PU/nHA wt.20% sample showed higher surface damage (fig. 4d) presenting more dissolution within paths or pores in the coating microstructure associated to the higher ceramic phase concentration.

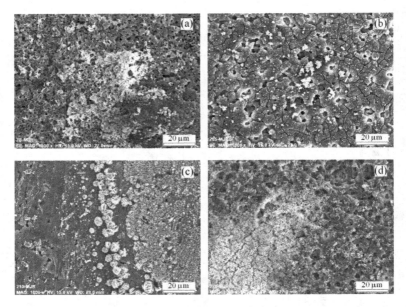

Figure 4. Images by SEM of: a) PU, b) PU/nHA wt.5%, c) PU/nHA wt.10%, y d) PU/ nHA wt.20% hybrid coatings after 24 hours of inmersion in Hank solution at 25 °C.

CONCLUSIONS

Corrosion studies and materials characterization shown hybrid coatings improve the corrosion resistance in Hank solution at 25 °C in static conditions, and allow the formation of small scale of corrosion products. In this way, PU/nHA wt.10% hybrid coating improved the higher corrosion resistance due lower paths and pores density in the coating microstructure and the coalescence between PU and nHA phases. Diffusion of ionic species was the governing mechanism in the corrosion behavior of hybrid coating.

Otherwise, the PU/nHA wt.10% hybrid coating showed higher electrochemical activity associated to the more electrolyte flux permeability thought the paths and pores of allowing the increase of PU biodegradation, in the same way, the growing of HA phase by the electrolyte interaction with nHA initial phase.

REFERENCES

1. J.B. Park, J.D. Bronzino, *"Biomaterials, principles and applications"* by CRC Press LLC, Boca Raton, Florida 33431, (2003)
2. Z.H. Baqain, W.Y. Moqbel, F.A. Sawair, *British J. of Oral and Maxillofacial Surgery*, **50**, 239 (2012).
3. V.A. Alves, R.Q. Reis, I.C.B. Santos, D.G. Souza, T. de F. Gonçalves, M.A. Pereira-da-Silva, A. Rossi, L.A. da Silva, *Corr. Sci.*, **51**, 2473 (2009).
4. A.A. Ghoneim, A.M. Fekry, M.A. Ameer, *Electroch. Acta*, **55**, 6028 (2010).
5. Y. Lai, Y. Li, L. Jiang, W. Xu, X. Lv, J. Li, Y. Liu, *J. Electroanal. Chem.*, **671**, 16 (2012).
6. A. Zomorodian, F. Brusciotti, A. Fernandes, M.J. Carmezim, T. Moura e Silva, J.C.S. Fernandes, M.F. Montemor, *Surf. Coat. Tech.* **206**, 4368 (2012).
7. D.C. Parada-Quinaya, H.A. Estupinan Duran, D.Y. Pena Ballesteros, C. Vasquez Quintero, D. Laverde Catano, *Ingeniare*, **17**, 365 (2009).
8. P. Vermette, H.J. Griesser, G. Laroche, R.Guidoin, "Biomedical applications of polyurethanes", Texas USA: Landes Bioscience Georgetown, (2001).
9. A.B. Martínez Valencia, G. Carbajal De la Torre, A. Duarte Moller, H.E. Esparza Ponce, M.A. Espinosa Medina, *Int. J. Phys. Sci.*, **6**, 6681 (2011).
10. A.B. Martinez Valencia, G. Carbajal De la Torre, R. Torres Sánchez, L. Téllez Jurado, H.E. Esparza Ponce, *Int. J. Phys. Sci.*, **6**, 2731 (2011).
11. G. Carvajal De La Torre, R. Nava Mendoza, M.A. Espinosa Medina, A. Martinez Villafañe, J.G. Gonzalez-Rodriguez, V.M. Castaño, *British Corr. J.*, **37**, 293 (2002).
12. G. Carbajal de la Torre, M.A. Espinosa Medina, A. Martinez Villafañe, J.G. Gonzalez Rodriguez, V.M. Castaño, *The Open Corr. J.*, **2**, 197 (2009).

Mater. Res. Soc. Symp. Proc. Vol. 1766 © 2015 Materials Research Society
DOI: 10.1557/opl.2015.424

Effect of Li-doping on Photoluminescence of Screen-printed Zinc Oxide Films

L. Khomenkova[1], V. Kushnirenko[1], M. Osipenok[1], K. Avramenko[1], Y. Polishchuk[1], I. Markevich[1], V. Strelchuk[1], V. Kladko[1], L. Borkovska[1] and T. Kryshtab[2]

[1] V. Lashkaryov Institute of Semiconductor Physics of NAS of Ukraine,
Pr. Nauky 41, 03028 Kyiv, Ukraine
E-mail: khomen@ukr.net

[2] IPN – ESFM, Av. IPN, Edificio 9 U.P.A.L.M. C.P. 07738 México
E-mail: kryshtab@gmail.com

ABSTRACT

Undoped and Li-doped ZnO films were fabricated by screen printing approach on sapphire substrate. The effect of Li doping and annealing temperature on the luminescent, optical, electrical and structural properties of the films has been investigated by the photoluminescence (PL), Raman scattering, conductivity, Atomic Force microscopy and X-ray diffraction (XRD) methods. The XRD study revealed that the films have polycrystalline wurtzite structure with grain sizes ranging from 26 to 38 nm. In the undoped ZnO films, the increase of annealing temperature from 800 to 1000 °C resulted in the increase of the grain sizes, film conductivity and the intensity of the ultraviolet PL. The introduction of Li of low concentration of 0.003 wt % at 800 °C or 900 °C allows producing the low-resistive films with enhanced ultraviolet PL and reduced density of crystalline defects. Highly doped films (with 0.3 wt % of Li) were found to be semi-insulating with deteriorated PL properties irrespectively of the annealing temperature. It is shown that introduction of Li in the ZnO films affects their PL spectra mainly via the evolution of the film crystallinity and the density of intrinsic defects.

INTRODUCTION

Over the past decades, ZnO-based materials have been attracted considerable attention for possible application in optoelectronic devices, especially to ultraviolet (UV) light emitters [1]. The structural, optical and electrical properties of ZnO films can be governed by dopants, deposition parameters and post-treatments. Doped ZnO films are of great interest for their applications in transparent conducting electrodes for solar cells and displays [2, 3] and insulating or ferroelectric layers for optical memory devices [4]. However, hardly achievable p-type conductivity of ZnO-related materials remains the drawback for their wide applications.

The large efforts were applied to obtain p-type ZnO by doping with elements of groups I (Li, Na, K) or V (N, P, As) as well as by their co-doping [1]. Among these, lithium (Li) has been considered as a promising dopant for ZnO [5-8] in spite of rather contradictory viewpoints on realization of stable p-type activity under equilibrium conditions.

As a rule, Li-doping occurs as follows [9]:

$$Li_2O \xrightarrow{ZnO} Li_{Zn}^- + Li_i^+ + O_O \qquad (1)$$

Where Li_{Zn}^- represents lithium on zinc lattice site, Li_i^+ is a lithium in interstitial position, and O_O is an oxygen on its lattice site.

The most of experimental reports supposed that Li_{Zn} is a deep acceptor with a binding energy of about 800 meV responsible for luminescence band peaked at 2.0-2.2 eV [10-13]. However, it was theoretically predicted that Li_{Zn} produces shallow acceptor level [14]. In fact, in some luminescent studies of Li-doped ZnO materials, the shallow acceptor states having binding energies of about 300 meV were demonstrated and ascribed to Li_{Zn} defects [15, 16]. These shallow acceptor states are responsible for luminescence band around 3.05 eV and are introduced only if the growth/diffusion temperature is below 700 °C, otherwise deep acceptor state (with the 800 meV binding energy) is formed [16]. The amphoteric behavior of Li in the ZnO together with an ability of Li_{Zn} acceptor to form neutral complexes with other defects has been suggested to be responsible for semi-insulating nature of ZnO:Li materials [4, 10].

Furthermore, if Li-doping is performed during ZnO crystallization, lithium can affect also the crystallinity of ZnO films. These data are also rather contradictory. Specifically, the X-ray diffraction (XRD) study of Li-doped ZnO films and ZnO powders has shown that Li introduction upon thermal annealing can suppress [17, 18] and/or promote [7, 8, 17, 18] the growth of nanocrystals in sizes. Besides, Raman scattering investigations of Li-doped ZnO revealed that Li can stimulate [19] or counteract [20, 21] the formation of point defects presumably the oxygen vacancies or Zn interstitials. However, the effect of Li on the film crystallinity usually is not taking into account when the role of Li_{Zn} acceptor in the photoluminescence (PL) of ZnO is considered.

In this work we have studied the effect of Li on the PL of ZnO thick films prepared by a screen printing method and doped with Li at different annealing temperatures. In addition, the influence of Li doping on the film crystallinity have been analyzed.

EXPERIMENTAL

The 40-μm films of undoped and Li-doped ZnO (ZnO: Li) were produced by the screen printing method on (1012) sapphire substrates of 10x10x0.5 mm³ dimension using ZnO-based paste (Fig.1). The paste was made from the ZnO powder (Sigma-Aldrich, 99.99%) crushed in a ball mill for 125 h and mixed with distillate water or with LiNO₃ aqueous solution.

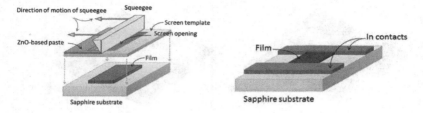

Figure 1. Schematic illustration of screen-printing technique used for ZnO films fabrication (left) and the sample with In contacts prepared by thermal evaporation with shadow mask (right).

The paste was distributed on the non-flexible template with opening places and then pushed towards a sapphire substrate using motion of the squeegee along the template. More details about fabrication approach can be found in [22]. After template removing, the films with required configuration were obtained on the substrate. As-printed films were dried at normal conditions for 24 h and then annealed in air at 800, 900 and 1000 °C for 30 min in muffle furnace. The Li concentrations in dried doped films were 0.003, 0.03 and 0.3wt %.

The PL and PL excitation (PLE) spectra were studied both at room and at liquid-nitrogen temperatures, whereas all other experiments were performed at room temperature only. The PL was excited by a 337.1 nm line of pulsed N_2-laser or by a light of Xe-lamp passed through a grating monochromatic. Atomic Force microscopy (AFM) was used to investigate the morphology of the films. The study of film surface was performed using NanoScope IIIa Dimension 3000TM (Digital Instruments). Measurements were carried out in contact mode using silicon probes (NT-MDT brand NSG01) with a tip radius of about 10 nm. The Raman scattering spectra were excited with a 488.0 nm line of an Ar-Kr ion laser and collected by a triple Raman spectrometer T-64000 Horiba Jobin-Yvon with a resolution about 0.15 cm^{-1}. XRD study was realized using X-ray powder diffractometer ARL X'TRA with the Cu $K\alpha_1$ and Cu $K\alpha_2$ radiation. For electric measurements, indium strip-like contacts were fabricated by thermal evaporation of pure In (99, 9%) onto the surface of ZnO film through shadow mask (Fig.1). Dark current-voltage characteristics were measured in the range of 1-40 V of applied bias, demonstrating linear behavior. To investigate photoelectric properties of the films, ultraviolet emission of a 500 W Hg lamp was used as excitation source.

RESULTS AND DISCUSSION

Morphology characterization

To obtain the information about sintering process of ZnO-based films, the surface morphology of the films was investigated by means of AFM method before and after annealing. Figure 2 represents three-dimensional AFM images obtained in tapping mode for as-printed dried film (Fig. 2a) and after film annealing at 1000 °C (Fig. 2b). It is seen that the surface of as-printed dried films consists of polycrystalline aggregates of ZnO grains with lateral dimensions in the range of 140-1800 nm and vertical dimensions within 50-1200 nm.

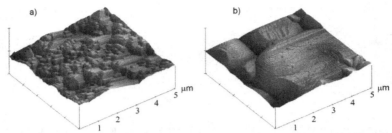

Figure 2. Three-dimensional AFM topography image of the surface of zinc oxide films obtained before (a) and after (b) annealing treatment at 1000 °C.

Smaller ZnO grains have fuzzy-like grain boundaries, whereas large grains demonstrate well organized layer-like structure. This is demonstrated by the appearance of the terraces with the height of 50-100 nm (Fig. 2a). After film sintering at 800-1000 °C, the recrystallization of the films was observed. As one can see from Fig. 2b, a significant increase of the external lateral and vertical dimensions of ZnO grain occurs. This process depends on the annealing conditions and for the samples annealed at 1000 °C, the grain size was found to be larger than 2 μm (Fig. 2b). In some cases the formation of the grains with the sizes up to 30 μm was observed (not shown here).

The Li-doped ZnO films sintering at different conditions were also investigated by AFM method. However, no significant difference was found in comparison with undoped films. The external dimensions of the ZnO grains were found to be in the range 0.5-2 μm that were similar to those obtained for undoped films, especially when the films were annealed at lower temperature. Based on this morphology observation, one can assume that during sintering the main transformation occurs inside large ZnO grains that has to affect optical, electrical and luminescent properties of the films.

Photoluminescence investigations

The room-temperature PL spectra of the ZnO and ZnO:Li films are shown in Figure 3. In the spectra, an UV PL band (I_{UV}) and a wide defect-related PL band (I_{DEF}) in the green-orange spectral range can be observed. Both the intensity and spectral position of these bands depend on the annealing temperature and Li concentration.

In the undoped ZnO films annealed at 800 °C and 900 °C, an intensity of the UV band is strongly reduced. It increases noticeably after the film annealing at 1000 °C. In the Li-doped films with the lowest Li concentration (0.003 wt%), strongly enhanced UV emission is observed already upon the annealing at 800 and 900 °C, but in the films with the highest Li concentration (0.3 wt%), no exciton emission is detected irrespective of the annealing temperature. In the last case the intensity of the visible defect-related band is found to be reduced too.

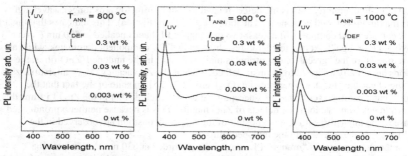

Figure 3. Room-temperature PL spectra of undoped and Li-doped ZnO films with different content of Li (0, 0.003, 0.03 and 0.3 wt %) annealed at 800 °C, 900 °C, 1000 °C for 30 min.

It should be noted, that in both undoped and Li-doped samples, spectral position of the I_{UV} band changes from 378 nm in the films with low intensity of this band to 384 nm in the films with strongly enhanced UV PL. In turn, spectral position of the maximum of the I_{DEF} band varied in the range of 522-555 nm in different samples and showed the shortest wavelength position in the ZnO:Li films doped with Li of 0.003 wt% and 0.03 wt% and annealed at 800 °C.

In the low-temperature PL spectra of the ZnO and ZnO:Li films (Fig. 4a) the same I_{UV} and I_{DEF} PL bands are present. Figure 4b shows the ratio of the intensity of the UV band to the intensity of the visible defect-related band that is usually considered for simple estimation of structural defects in ZnO. Similarly to room-temperature PL spectra, the best PL characteristics are observed in Li-doped films with low Li concentration, while the worth PL properties are found for highly doped ZnO:Li films (no UV emission). An enhancement of the relative intensity of the UV PL caused by Li introduction is observed: in the films annealed at 800 °C for 0.003 and 0.03 wt % of Li, in the films annealed at 900 °C for 0.003 wt % of Li only, while in the films annealed at 1000 °C no improvement of the UV PL is found.

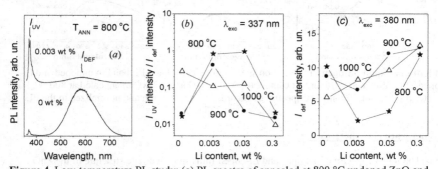

Figure 4. Low-temperature PL study: (a) PL spectra of annealed at 800 °C undoped ZnO and ZnO:Li (0.003 wt %) films under N_2-laser excitation; (b) Ratio of integrated intensities of the I_{UV} to the I_{DEF} band versus Li content; (c) Intensity of the I_{DEF} band under defect-related excitation versus Li content.

At low-temperature, the visible defect-related PL band shifts towards longer wavelengths and its maximum is found to be peaked at about 580-590 nm (Fig. 4a). This indicates that this PL band is complex and composes of at least two overlapped PL bands peaked at 520 nm ("green") and 590 nm ("orange"). At low temperatures, the "orange" PL component dominates, while at room temperature the "green" one contributes mainly to I_{DEF}. In the undoped ZnO films, these PL bands are usually assigned to intrinsic defects, such as the Zn vacancy (V_{Zn}), interstitial zinc (Zn_i), the O vacancy (V_O), oxygen antisite (O_{Zn}), etc. [1,23-25]. In spite of the fact that the "orange" band is often ascribed to Li_{Zn} acceptor we cannot exclude that in the ZnO:Li films this band originates from the intrinsic defects of ZnO mainly. This is because neither a pronounced increase in the intensity, nor the evident shift of the I_{DEF} peak position are observed in the ZnO:Li films as compared with the undoped films.

In the PLE spectra of "orange" PL component (recorded at 630 nm to eliminate the contribution of "green" PL component), in addition to a small peak at 369 nm corresponding to generation of exciton, a wide maximum at 380-385 nm is found (Fig. 5). The maximum at 385 nm is not observed in the PLE spectra of "green" PL component (recorded at 520 nm) and it is apparently caused by some defect-related absorption.

In the ZnO films annealed at 800 and 900 °C as well as in all ZnO:Li films with the highest Li content, the intensity of "orange" PL component under defect-related excitation (λ_{exc}=380 nm) exceeds that under band-to-band excitation. Figure 4c shows that, in general, in the ZnO:Li films, the intensity of "orange" PL band under defect-related excitation increases both with the increase of Li concentration and the annealing temperature. This means that introduction of Li stimulates the formation of intrinsic defects or their complexes with Li ions.

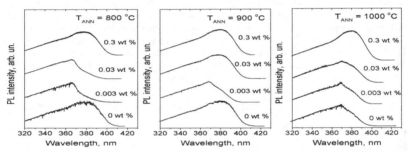

Figure 5. Liquid-nitrogen PL excitation spectra of defect-related PL band in undoped and Li-doped ZnO films with different Li content (0, 0.003, 0.03 and 0.3 wt %) annealed at 800 °C, 900 °C, 1000 °C for 30 min. The spectra are recorded at λ_{PL}=630 nm and normalized to the PL intensity at 337 nm.

Raman scattering spectroscopy

The Raman spectra of undoped and Li-doped films show wurtzite structure of the ZnO (Fig. 6). By comparison with the assignments of pure ZnO, Raman peaks at 99, 332, 379, 410 and 438 cm⁻¹ correspond to E_2^{low}, E_2^{high}-E_2^{low}, $A_1(TO)$, $E_1(TO)$ and E_2^{high} modes, respectively [26].

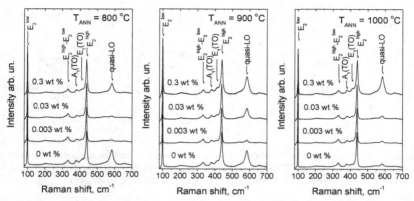

Figure 6. Room-temperature Raman spectra of undoped and Li-doped ZnO films with different content of Li (0, 0.003, 0.03 and 0.3 wt %) annealed at 800 °C, 900 °C, 1000 °C for 30 min. Raman active modes are indicated. The spectra are normalized on the intensity of E_2^{high} mode. λexc=488.0 nm.

The peak at 584 cm^{-1} marked as a quasi-LO mode lies at an intermediate frequency between the A_1(LO) and E_1(LO) modes and is ascribed to a quasi-mode of mixed A_1 and E_1 symmetry. Both the change in the annealing temperature and Li-doping affect the characteristics of E_2^{high} mode associated with oxygen-atom vibrations in ZnO lattice (Fig. 7a, b). Specifically, in the undoped ZnO films annealed at 800 and 900 °C as well as in highly doped ZnO films, the E_2^{high} mode is found to be broadened and shifted towards higher frequencies (up to 439 cm^{-1}) as compared with the bulk value of 437 cm^{-1}. This testifies to a decrease of lattice parameters in ZnO nanocrystals, which can be caused by external compressive stress, the presence of vacancy-related defects or Li incorporation in the Zn sublattice.

In Raman spectra of ZnO:Li films with 0.003 wt % of Li annealed at 800 °C and 900 °C, a low frequency shift and a decrease of a full width at a half maximum (FWHM) of the E_2^{high} mode are observed indicating strain relaxation and improvement of oxygen sublattice crystal structure (Fig. 7b). Further increase of Li concentration produces a reverse trend in the spectral shift and FWHM of the E_2^{high} mode.

The appearance of the quasi-LO mode in the Raman spectra has been attributed to the structural disorder in the ZnO such as V_O, Zn_i or their complexes [27]. Figure 7c shows that in the undoped ZnO films annealed at 800 and 900 °C, intensity of the quasi-LO mode is high and reduces in the film annealed at 1000 °C. This indicates the decrease of structural disorder in the films annealed at higher temperatures. At the same time, in the ZnO:Li films with lower Li concentration, the quasi-LO mode become weak that testifies to the decrease of density of structural defects owing to Li incorporation. However, all films with the highest Li concentration demonstrate enhanced quasi-LO mode. This means that introduction of high amount of Li produces additional structural disordering.

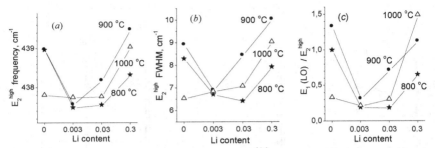

Figure 7. Peak position (a) and FWHM (b) of the E_2^{high} mode as well as integrated intensity ratio of the quasi-LO and E_2^{high} modes (c) versus Li content in the film annealed at different temperatures: 800 °C (stars), 900 °C (circles), 1000 °C (triangles).

Photo- and dark conductivity investigation

The results of conductivity investigation in the undoped and Li-doped films are summarized in Figure 8. The lowest conductivity was found in the undoped films annealed at 800 and 900 °C as well as in all ZnO:Li films with the highest Li content. In turn, the ZnO film annealed at 1000 °C as well as the ZnO:Li films with the lowest Li content (0.003 wt%) demonstrate a hundredfold larger dark and photo-conductivities.

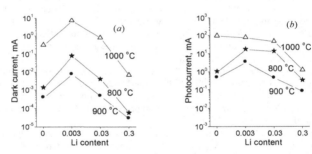

Figure 8. Room-temperature dark (a) and photo- (b) conductivities of undoped and Li-doped ZnO films annealed at 800 °C (stars), 900 °C (circles), 1000 °C (triangles) versus Li content. U=9 V.

X-ray diffraction study

The XRD patterns of undoped and Li-doped ZnO films show the peaks corresponded to polycrystalline wurtzite ZnO without preferable grain orientation (Fig. 9a). The marked extra peaks are due to the contacts. In the undoped films, the increase of annealing temperature leads to the shift of XRD peaks towards the lower angles and the decrease of peak FWHM indicating the increase of ZnO lattice parameters and grain size, respectively (Fig. 9a). The same effect is observed in the ZnO:Li film with the lowest Li content annealed at 800 °C. At the same time for ZnO:Li films with the highest Li content the diffraction peaks shift towards larger angles and FWHM increases. Assuming a homogeneous strain across the films, the crystallite sizes were evaluated from FWHM for different XRD peaks using the Scherrer equation. The representative behavior of grain sizes in dependence on annealing temperature and Li content is shown in Figure 9b for (002) diffraction peak.

In the undoped ZnO films, the crystallite sizes increase from 26 to 38 nm when the annealing temperature increases. In polycrystalline ZnO films, it is assumed that at higher temperatures the atoms have enough activation energy to occupy the proper sites in the crystal lattice, and larger grains with lower surface energy are formed [28]. The introduction of Li of low concentration also promotes nanocrystal growth at low temperatures while the highest Li content hinder this process (Fig.9b). The nonmonotonic dependence of the ZnO grain sizes on Li content in the ZnO:Li films has been observed also in [8, 17]. It has been supposed that diffusivity of the interstitial zinc, which is considered to play an important role in the grain growth of ZnO, is higher in the Li-doped ZnO than in the undoped one, because Li^+ has a smaller ionic radius of 0.060 nm than Zn^{2+} having 0.074 nm [8]. This can explain the increase of grain size in the films with low Li content. The decrease of grain size in the films with the highest Li content can be caused by segregation of insoluble Li atoms at the grain boundaries, which suppress the growth of nanocrystals [18]. In fact, the redundant Li elements congregated as the second phases on the grain boundaries of ZnO nanocrystals has been revealed by high resolution X-ray photoelectron microscopy in Li-doped ZnO powders with Li content of about 5 at % [7]. The dependence presented in Fig. 9b indicates that not only Li content but also the annealing temperature is important for the ultimate effect of Li doping on the grain growth.

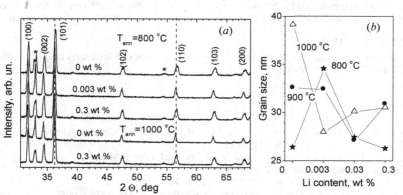

Figure 9. XRD patterns (a) and grain sizes (b) of undoped and Li-doped ZnO films vs Li concentration and annealing temperature. The peaks marked with * are caused by contact.

The increase of crystal sizes is accompanied by the improvement of PL and optical characteristics as well as by the increase of film conductivity and vice versa, although this interconnection is not so direct. The increase of crystal sizes decreases the density of intercrystalline boundaries, which act as the sinks for point defects, in particular, for oxygen ions. The oxygen ions chemisorbed at the film surface and at the grain boundaries are known to be the origin of low dark conductivity in undoped polycrystalline ZnO films [29]. Therefore, higher dark conductivity in the films with larger crystal sizes is seemingly connected with the decrease of extended defects, though the appearance of Li_i donors in the ZnO:Li films can increase conductivity too. A fairly good correlation of the PL characteristics (relative intensity of the UV PL, the appearance of the defect-related maximum in the PLE spectra) and optical properties (the characteristics of the E_2^{high} mode, the appearance of the quasi-LO mode in the Raman spectra) is found. This allows supposing that modification of the ZnO films microstructure caused by Li introduction affects also the formation of vacancy and interstitial defects inside the crystals as well as stimulate the defect diffusion during crystal growth.

CONCLUSIONS

The present work demonstrates the utility of screen printing technique for the fabrication of undoped and Li-doped ZnO films. The films annealed at 800-1000 °C were found to be polycrystalline of wurtzite structure. The undoped ZnO films showed the increase of the grain sizes and the decrease of the concentration of crystalline defects with the increase of the annealing temperature. The improvement of the films crystallinity was accompanied by an increase of the intensity of the UV PL and both dark- and photo-conductivities. The effect of Li on the films crystallinity as well as on the PL and electrical properties was found to depend not only on Li content but also on the annealing temperature. The doping with low Li concentration (0.003 wt%) at 800 °C allowed producing the films with improved characteristics and thus decreasing thermal budget for film production. On the contrary, highly doped films (0.3 wt% of Li) were found to be semi-insulating, having smaller crystal sizes and a lot of crystalline defects that result in dramatic quenching of the UV PL. It is shown that the effect of Li-doping on both UV and visible defect-related PL bands appears mainly in the influence of Li on the development of film microstructure via the control of crystal sizes and concentration of native point defects.

ACKNOWLEDGMENTS

This work was supported by National Academy of Sciences of Ukraine via the project III-10-12 and by CONACYT, Mexico.

REFERENCES

1. Ü. Özgür, Ya. I. Alivov, C. Liu, A. Teke, M.A. Reshchikov, S. Doğan, V. Avrutin, S.J. Cho, H. Morkoç, *J. Appl. Phys.*, **98**, 041301 (2005).
2. T. Minami, H. Nanto, S. Takata, *Jpn. J. Appl. Phys.*, **23**, L280 (1984).
3. J. Hu, R.G. Gordon, *Solar Cells* **30**, 437 (1991).
4. M. Joseph, H. Tabata, T. Kawai, *Appl. Phys. Lett.*, **74**, 2534 (1999).
5. S.H. Jeong, B.N. Park, S.B. Lee, J.H. Boo, *Thin Solid Films*, **516**, 5586 (2008).
6. M. Caglar, Y. Caglar, S. Aksoy, S. Ilican, *Appl. Surf. Sci.*, **256**, 4966 (2010).
7. B. Wang, L. Tang, J. Qi, H. Du, Zh. Zhang, *J. Alloys Compd.*, **503**, 436 (2010).
8. Sh. Fujihara, Ch. Sasaki, T. Kimura, *J. Eur. Ceram. Soc.*, **21**, 2109 (2001).
9. P. Bonasewicz, W. Hirschwald, G. Neumann, *J. Electrochem. Soc.*, **133**, 2270 (1986).
10. W. Water, S.Y. Chu, Y.D. Juang, S.J. Wu, *Mater. Lett.*, **57**, 98 (2002).

11. C. Rauch,W. Gehlhoff, M.R. Wagner, E. Malguth, G. Callsen, R. Kirste, B. Salameh, A. Hoffmann, S. Polarz, Y. Aksu, M. Driess, *J. Appl. Phys.*, **107**, 024311 (2010).
12. V.I. Kushnirenko, I.V. Markevich, T.V. Zashivailo, *J. Lumin* **132**, 1953 (2012).
13. N. Ohashi, N. Ebisawa, T. Sekiguchi, I. Sakaguchi, Yo Wada, *Appl. Phys. Lett.*, **86**, 091902 (2005).
14. M.G. Wardle, J.P. Goss, P.R. Briddon, *Phys. Rev. B*, **71**, 15520 (2005).
15. Z. Zhang, K.E. Knutsen, T. Merz, A.Yu. Kuznetsov, B.G. Svensson, L.J. Brillson, Appl. Phys. Lett., **100**, 042107 (2012).
16. B.K. Meyer, J. Stehr, A. Hofstaetter, N. Volbers, A. Zeuner, J. Sann, *Appl. Phys. A*, **88**, 119 (2007).
17. D.Y. Wang, J. Zhou, G.Z. Liu, *J. Alloys Compd.*, **481**, 802 (2009).
18. P. Chand, A. Gaura, A. Kumar, U.K. Gaur, *Ceramics Intern.* **40**, 11915 (2014).
19. Y.L. Du, Y. Deng, M.S. Zhang, *Sol. St. Commun*, **137**, 78 (2006).
20. R. Yousefi, A. Khorsand.Zak, F. Jamali-Sheini, *Ceramics Intern*, **39**, 1371 (2013).
21. C. Bundesmann, N. Ashkenov, M. Schubert, D. Spemann, T. Butz, E.M. Kaidashev, M. Lorenz, M. Grundmann, *Appl. Phys. Lett.*, **83**, 1974 (2003).
22. M. Osipyunok, G. Pekar, O. Syngaivskyy, *The method of fabrication of solid layers by screen printed approach,* Patent of Ukraine 94561 (10.05.2011).
23. C. Ton That, L. Weston, M.R. Phillips, *Phys. Rev. B*, **86**, 115205 (2012).
24. A.F. Kohan, G. Ceder, D. Morgan, *Phys. Rev. B*, **61**, 15019 (2000).
25. M. Liu, A.H. Kitai, P. Mascher, *J. Lumin.*, **54**, 35 (1992).
26. J.M. Calleja, M. Cardona, *Phys. Rev. B*, **16**, 3753 (1977).
27. J.N. Zeng, J.K. Low, Z.M. Ren, T. Liew, Y.F. Lu, *Appl. Surf. Sci.*, **197**, 362 (2002).
28. Z.B. Fang, Z.J. Yan, Y.S. Tan, X.Q. Liu, Y.Y. Wang, *Appl. Surf. Sci.*, **241**, 303 (2005).
29. S.A. Studenikin, N. Golego, M. Cocivera, *J. Appl. Phys.*, **87**, 2413 (2000).

Characterization of Nanostructured Materials

Mater. Res. Soc. Symp. Proc. Vol. 1766 © 2015 Materials Research Society
DOI: 10.1557/opl.2015.425

Synthesis of AlFe Intermetallic Nanoparticles by High-Energy Ball Milling

G. Rosas[1*], J. Chihuaque[1], E. Bedolla[1], R. Esparza[2] and R. Pérez[2]

[1] Instituto de Investigaciones Metalúrgicas, UMSNH, Edificio U, Ciudad Universitaria, Morelia, Mich., 58000, México. *Email: grtrejo@umich.mx

[2] Centro de Física Aplicada y Tecnología Avanzada, Universidad Nacional Autónoma de México, Boulevard Juriquilla 3001, Santiago de Querétaro, Qro., 76230, México.

ABSTRACT

In this investigation, the chemical and microstructural characteristics of nanostructured AlFe intermetallic produced by high-energy ball milling have been explored. High purity elemental powders were used as the starting material. The ball milling was carried out at room temperature using a SPEX-8000 mixer/mill. The structure, morphology and compositions of the powders were obtained using X-ray diffraction patterns (XRD), scanning and transmission electron microscopy (STEM). High resolution electron microscopy observations have been used in the nanostructured materials characterization. The structural configurations have been explored through comparisons between experimental HREM images and theoretically simulated images obtained with the multislice method of the dynamical theory of electron diffraction.

INTRODUCTION

Nanoscale materials have stimulated great interest due to their importance in basic scientific research and potential technological applications, which exhibit unique chemical and physical properties, differing substantially from the bulk counterpart [1-2]. The synthesis of nanostructured materials is one of the main research topics in the nanoscience scientific and technological community [3]. Since the discovery of the carbon, nanotubes by Iijima [4] different methods have been explored to produce nanomaterials [5, 6]. Thus, for example, chemical methods have been commonly used for the synthesis of nanoparticles, nanotubes and nanorods [7, 8]. However, other procedures have also been successful used [9]. Mechanosynthesis has been recently used as a procedure to synthesized nanostructures [10]. Mechanosynthesis processes are known as: mechanical milling, mechanical alloying and reactive milling. Mechanical milling is referring to the milling of a pure metal or compound. Mechanical alloying is referring to the formation of alloys from elemental precursors during the process. Reactive milling uses mechanical processing to induce chemical reactions [11]. Also, mechanosynthesis has been considered as a route of green Chemistry processes [12]. In recent years, the mechanochemical processes have been used for the synthesis of different metallic nanoparticles, nanorods and nanotubes [13, 14]. The magnetic properties of nanostructured iron materials have drawn special attention due to potential applications in sensors and storage devices. Also, iron has been found to be one of the main catalytic elements for the growth of nanotubes [15]. In order to know the mechanism of formation of these structures it is important to study their mechanism of growth starting from simple elements, for instance metallic nanoparticles. Metallic nanoparticles have very interesting properties in catalysis and some are potential candidates for hydrogen storage and others are used in industrial processes involving gas treatments [16]. Finally, due to their particular confined electronic configuration they can also be used as conducting elements in building of nanostructures, such as nanotubes and nanorods [17]. Intermetallic compounds are made of two or more metals or of a metal and a nonmetal. Aluminides of transition metals such as iron, nickel, titanium, and cobalt are a few examples. Intermetallic compounds of stoichiometric composition generally assume an ordered crystal structure. Therefore, different types of nanoparticles made of intermetallic compounds are being investigated as they can be used in industrial applications such as ultra-high-density magnetic storage systems [18, 19]. In this investigation, the chemical and microstructural characteristics of different nanostructured materials like intermetallic nanoparticles and carbon nanotubes produced by mechanosynthesis processes have been explored.

EXPERIMENTAL

High-energy mechanical milled compounds were obtained using elemental powders such as Al (99.99%) and Fe (99.99 %). All the mechanical milled reactions were carried out at room temperature in a high-energy vibratory mill SPEX 8000 mixer/mill with a milling speed of 1800 rpm using hardened steel balls and steel vials. The used steel balls were of 8.27 g and 12.7 mm in diameter. An argon atmosphere was used during the milling process. Different milling time and a 12:1 ball to powder relation were used to each mechanical milled compound. Ethanol was added to the milling mixture to avoid the sticking of the powders to the vial walls. The nanomaterials mechanically milled obtained powders were structurally characterized using X-ray diffractometry (Siemens D5000, CuKα radiation) and transmission electron microscopy (FEG-TEM Philips Tecnai F20). The TEM observations were carried out using powder samples deposited on carbon-coated copper grids. HREM simulated images have been obtained using the electron diffraction dynamical theory (multislice approach) and some insights on the structural characteristic of nanometric metallic particles and nanotubes are withdrawn.

RESULTS AND DISCUSSION

AlFe Nanoparticles

AlFe intermetallic powders have been obtained using the mechanical alloying technique with different milling periods (3, 6, 10 and 14 hours). The chemical structural nature of the alloyed powders is obtained from X-ray diffraction patterns. A series of typical X-ray diffraction patterns are shown in Fig. 1a. The spectrum labeled "0 h" shows the initial Al, Fe elemental powders. After a milling period of 3 hours, traces of the AlFe intermetallic compound can clearly be distinguished. Considering a milling period of 14 hours, the indications of the AlFe presence is more evident. The X-ray spectrum peaks are wide in nature suggesting indeed a small size (nanometric) of the AlFe crystallites. This is better understood with the help of the theoretical X-ray diffraction patterns shown in Fig. 1b, which shows the patterns obtained from the crystalline specimen sizes of 50 nm and for crystallites in the range of 1.5 nm. The broadening and the intensity decrease of the theoretical peaks are clearly illustrated. These theoretical patterns resemble the experimental pattern profiles and denote the existence of two different scales of clusters, the micrometer size particle powders and the nanometric AlFe clusters with sizes around 2–4 nm.

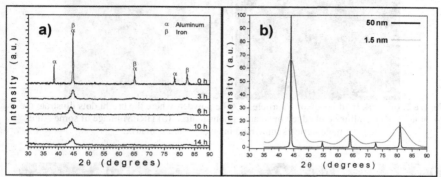

Figure 1. (a) X-ray diffraction pattern obtained after different mechanical alloyed periods. After 6 hours the spectra shows evidence of the AlFe intermetallic compound. (b) Theoretical calculations of X-ray diffraction patterns for AlFe with crystals sizes of 50 nm and 1.5 nm.

There is a strong dependence of the average particle size as a function of the mechanical alloyed milling time. After 3 milling hours, the average particle size is of the order of 100 nm. However, after 14 milling hours, the average particle size is approximately 73 nm. Therefore, there is a clear tendency of crystal size diminishing as a function of the increments of the milling time. Furthermore and in accordance with the parameters observed previously by X-ray diffraction, it has also been found that these particle powders are formed of nanometric size clusters embedded in an amorphous matrix. Evidence of this result is shown in Fig. 2a, where it can be observed a high angle annular dark field image (which produces a Z equivalent contrast) obtained from an aggregate of the AlFe compound. This image shows a large number of bright dots in the nanometric size range, which must correspond to small metallic clusters inside a matrix with a lower Z number. In order to understand these clusters, high resolution electron microscope (HREM) images have been obtained from these nanometric particles. This is illustrated in Fig. 2b. In the HREM image, it is possible to distinguish some cubic-like and multiple-twinned nanoparticles of approximately 2 nm immersed in an amorphous matrix. In fact, this allows recognizing that the global morphology involves aggregates of around 100 nm. However, the formation of AlFe metallic nanocrystals is in the order of just a few nanometers.

It is interesting to mention that perfect cubic particles can hardly be found. Most of the nanometric particles are cubic-like deformed particles. This is illustrated in Fig. 3 where nanometric cubic particles HREM images are displayed. The deformed nature of these particles can clearly be seen in Figs. 3b, 3c and 3d. Another interesting image contrast feature showed in the HREM images of the nanometric particles is related, for example, with the presence of brighter dots in these HREM images in comparison with the other bright dots of the total image. These effects are indicated by arrows.

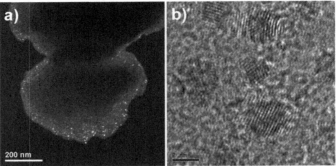

Figure 2. (a) Dark field image of a particle of AlFe powder, where the bright dots illustrate the nanometric crystallites embedded in an amorphous matrix. (b) HREM image of some AlFe nanoparticles immersed in an amorphous matrix.

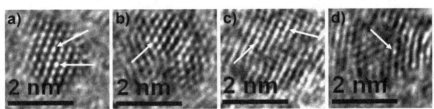

Figure 3. HREM images of AlFe nanoparticles showing brighter dots in the overall contrast (signaled with arrows), (a) cubic-perfect nanoparticle, (b), (c) and (d) cubic-deformed nanostuctures.

Trying to understand the nature of the brighter dots, Fig. 4 shows HREM image simulations along the [001] and [111] zone axis of an AlFe intermetallic cluster, where a random number of Fe vacancies have been induced. The presence of these vacancies give rise to brighter image contrast dots in the HREM images (indicated by arrows). Also, the intensity profile shows that the intensity profile is not homogeneous, where peaks with more intensity that others ones can be observed. This effect qualitatively resembles the experimental obtained results illustrated in Fig. 3. Therefore most of the small metallic particles obtained from mechanically alloyed elemental powders are deformed in nature and possible with the presence of Fe vacancies.

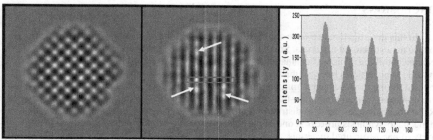

Figure 4. HREM theoretical simulations of an AlFe cluster with a random distribution of Fe vacancies along the [001] and [101] zone axis orientation with their intensity profile.

Carbon Nanotubes and Nanoparticles of AlFe

Carbon nanotubes and small metallic particles (1-3 nm) of AlFe have been obtained using mechanical alloying technique. The main source of the carbon present in these alloyed components is ethanol used to avoid the agglomeration. Figs. 5a, 5b and 5c show transmission electron microscopy images of the nanotubes. These nanotubes are commonly a few hundred nanometer long and approximately 40 nm wide. In a low magnification image (Fig. 5b), we can identify straight nanotubes, however, most of the formed nanotubes follow random trajectories with an aluminum-iron particle at one of their ending extremes. The production of these nanotubes is induced by the energy of the milling process and with the apparent help of AlFe clusters as catalysts. Fig. 5c clearly illustrates the multilayer nature of these nanotubes. One of the most interesting morphological aspects of these multilayer nanotubes is the presence of particles at the nanotube ending extremes. The compositional analysis of these particles showed that they are iron rich.

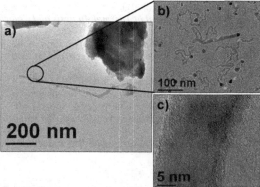

Figure 5. (a) Low magnification bright field image of a nanotube, (b) Bright field image showing different nanotubes with an AlFe particle at one of their ends, (c) HREM image of a nanotube where the multilayer nature is clearly illustrated.

CONCLUSIONS

In this investigation, we have shown that mechano-synthesis processes can be used to produce nanostructured materials. Small nanoparticles of AlFe are formed after 6 hours of mechanical milling. This was evaluated experimentally by HAADF-STEM images. The produced nanoparticles have a cubic-deformed structure which must be related with energy provided during the mechanical milling process. Carbon nanotubes with random trajectories were produced with an aluminum-iron nanoparticle at one of their ending extremes. Simulated HREM images suggest random distribution of Fe vacancies. The previous examples show that the mechano-synthesis process is one of the most interesting technologies to produce a great variety of nanostructured materials.

REFERENCES

1. P. Moriarty, *Rep. Prog. Phys.*, **64**, 297 (2001).
2. E. Guihen, J. D. Glennon, *Analytical Letters*, **36**, 3309 (2003).
3. A.G. Mamalis, *Journal of Materials Processing Technology*, **181**, 52 (2007).
4. S. Iijima, T. Ichihashi, *Nature*, **363**, 603 (1993).
5. X. Hu, W. Cheng, T. Wang, E. Wang, S. Dong, *Nanotechnology*, **16**, 2164 (2005).
6. Y. Wang, M.J. Kim, H. Shan, C. Kittrell, H. Fan, L.M. Ericson, W. Hwang, S. Arepalli, R.H. Hauge, R.E. Smalley, *Nano Lett.*, **5**, 997 (2005).
7. S. Panigrahi, S. Kundu, S. K. Ghosh, S. Nath, T. Pal, *Journal of Nanoparticle Research*, **6**, 411 (2004).
8. R. Esparza, J.A. Ascencio, G. Rosas, J.F. Sánchez Ramírez, U. Pal, R. Perez, *Journal of Nanoscience and Nanotechnology*, **5**, 641 (2005).
9. M.C. Daniel, D. Astruc, *Chem. Rev.*, **104**, 293 (2004).
10. M.K. Beyer, H. Clausen Schaumann, *Chem. Rev.*, **105**, 2921 (2005).
11. C. Suryanarayana, *Prog. Mater. Sci*, **46**, 1 (2001).
12. K. Wieczorek Ciurowa, K. Gamrat, *Journal of Thermal Analysis and Calorimetry*, **88**, 213 (2007).
13. G. Rosas, J. Sistos, J.A. Ascencio, A. Medina, R. Pérez, *Appl. Phys. A*, **80**, 377 (2005).
14. J. Ayala-Sistos, G. Rosas, R. Esparza, R. Pérez, *Adv. in Tech. of Mat. and Mat. Proc. J. (ATM)*, **7**, 175 (2005).
15. K.B. Kouravelou, S.V. Sotirchos, *Rev. Adv. Mater. Sci.*, **10**, 243 (2005).
16. Q.Q. Zhao, A. Boxman, U. Chowdhry, *Journal of Nanoparticle Research*, **5**, 567 (2003).
17. G. Schmid, *Nanoscale materials in chemistry*, John Wiley & Sons, (2001).
18. B. Rellinghaus, S. Stappert, M. Acet, E.F. Wassermann, *Journal of Magnetism and Magnetic Materials*, **162**, 142 (2003).
19. S. Stappert, B. Rellinghaus, M. Acet, E.F. Wasseramann, *The European Physical Journal D*, **24**, 351 (2003).

AUTHOR INDEX

SUBJECT INDEX